# JOHN AUBREY
## *and the* ADVANCEMENT *of* LEARNING

Incipit prologus in librum qui Flores hystoriarum intitulatur.

Temporum summam lineamque descendentem ab exordio mundi vel a successionibus quorundam regnorum et regum ad erudicionem futurorum diutinam annotare ut ex breuibus lector deuotus plura colligere possit. Sed quia quosdam auditores pigros dicentes qui obiectando dicunt quid necesse est aut vitas vel mortes seu diuersos hominum casus litteris mandare, prodigia celi et terre vel aliorum elementorum scriptis ipsa perpetuare. Nouerit iste bona vita et mores precedentium ad imitacionem sibi sequentium prouocant, malorum vero exempla ut non imitentur deterrent, prodigia autem vel portenta preterita que famem vel mortalitatem vel aliqua similia indicarent flagella pro meritis filiorum hominum cum videntur significant ideo memorie pro litteris commendantur ut si quando similia euenerint peccatores qui se iram dei in aliquo incurrisse meminerint mox ad remedium penitencie et confessionis pro dei placacione festinent. Ob hanc itaque causam viri alii non desunt Moyses legislator in hystoria diuina innocenciam Abel, Caym inuidiam, simplicitatem Iacob, dolositatem Esau, maliciam iudeorum filiorum Israel, bonitatem duodecim Ioseph, penam quinque ciuitatum pro consumpcione ignis et sulphuris manifestat, quatinus bonos imitemur, malorum esse sequaces abhorreamus, ut penam ponentes fugiendo vadamus. Quem et in non solum Moyses sed omnes diuine pagine tractatores in hystoricis et in moralibus libris faciunt virtutes amando vicia detestando deum timere simul et amare nos admonendo. Non est ergo audiendus qui cronicorum libros et martyrum a catholicis editos dicit esse negligendos, pro quibus eque humane sapiencie et saluti necessarium est per memoriam inuenire, per intelligenciam cognoscere, et per facundiam proferre studiosus valeat.

Modo igitur ab opere primordiali sex dierum. Sex diebus deus rerum creaturam formauit. Primo condidit lucem. Secundo firmamentum celi in medio librauit aquarum, superiores aquas ac terram cum celo superiore continentibus que in ea condiore laudarent. Tercio die eiusdem exordium creatis. Tercio congregatis in suum locum aquis que aridam cooperiebant aridam iussit apparere. Quarto sydera in firmamento celi posuit quem nunc quintum pro coniecturam equinoctii colligitur xii kl. apr. dies vocatur. Quinto natatilia volatilia et animancia creauit. Sexto alia terrestria et primum hominem Adam in agro Damasceno formauit et de eius latere dormientis Euam omnium viuencium matrem produxit, qui nunc quintum decimum credibile videtur x kl. aprilium dies appellatur. Unde multo credit si non velut veriorum sentenciam vincit eodem x kalendas aprilium die dominum fuisse crucifixum. Decebat enim ut vna eademque non solum ebdomadis sed et mensis die secundus Adam pro genere humani salute mistica morte sopitus produceret ex latere suo celestibus sacramentis sponsam scilicet ecclesiam omnium, quo videlicet die primi

# JOHN AUBREY
## *and the* ADVANCEMENT *of* LEARNING

William Poole

Bodleian Library
UNIVERSITY OF OXFORD

*Frontispiece* The *Flores Historiarum* of 'Matthew of Westminster' (a ghost; the author is really Matthew of Paris), an example of a (now rather battered) medieval manuscript gifted by Aubrey to the Bodleian. Oxford, Bodleian Library, MS e Musæo 149.

First published in 2010 by the Bodleian Library
Broad Street
Oxford OX1 3BG

www.bodleianbookshop.co.uk

ISBN 978 1 85124 319 8

The Publisher gratefully acknowledges the support of 'Cultures of Knowledge: An Intellectual Geography of the Seventeenth-Century Republic of Letters', a collaboration between the Bodleian Library and the Humanities Division of the University of Oxford with generous funding from The Andrew W. Mellon Foundation, and the generous assistance of Gene and Carol Ludwig.

Designed by Dot Little
Typeset in Adobe Caslon 10.5 on 14 pt body text, with display headings in Tagettes
Printed and bound by Great Wall Printing, China
British Library Catalogue in Publishing Data
A CIP record of this publication is available from the British Library

# Preface and Acknowledgements

This short book is intended as an accessible introduction to the intellectual world of the seventeenth-century natural philosopher and antiquary John Aubrey (1626–1697). It has been written to accompany the parallel exhibition held by the Bodleian Library in the summer of 2010. The year 2010 also marks the 350th anniversary of the Royal Society of London; Aubrey's relation to that club, of which he was a founding fellow, and the relation of the Royal Society itself to earlier Oxford initiatives, are therefore major preoccupations of this book. For those readers who attend the exhibition, I should remark that each chapter parallels one case in the accompanying exhibition, and discusses the context of its contents, although here as part of a sustained narrative. Most of the chapters are short, but slightly more attention has been given to the areas in which Aubrey's participation seems especially significant, notably the study of megaliths and the construction of an artificial language. I hope the reader will forgive the occasional repetition, so that each chapter might be rendered coherent in itself. Aubrey himself is the centre of the book, but he is best understood as a participant in various larger groups of scholars. It is therefore Aubrey's intellectual world that is the subject of this book, and not Aubrey in isolation. Aubrey was a very clubbable man in a very clubbable century.

As befits an introductory work, this book is largely a synthesis of accepted scholarship on the issues I discuss, and it is right for me to record my deep indebtedness to Michael Hunter for his pioneering work on all of Aubrey's intellectual life; to Kate Bennett for her expertise on Aubrey's biography and biographical work; to Mordechai Feingold for his scholarship on the history of higher education and the origins of learned societies; to Rhodri Lewis for his definitive book on artificial languages; and to Jim Bennett and Stephen Johnston for their work on the history of their own institution, the Old Ashmolean Museum, now the Museum of the History of Science. I summarise, too, some material from my own *The World Makers* (Peter Lang, 2010). At the Bodleian, among many others, Richard Ovenden, Dana Josephson and Madeline Slaven must stand out for their help with the exhibition which this book accompanies, and among the publishers I am particularly grateful to Samuel Fanous and Deborah Susman, and to Dot Little for designing this book. Finally, Rhodri Lewis and Kate Bennett read through the manuscript at very short notice and saved me from several errors of fact or emphasis.

William Poole, New College, October 2009

# I

## *John Aubrey, Virtuoso*

Today, we would not be able to find one word to classify the rich intellectual life of the seventeenth-century English polymath John Aubrey (1626–1697). We would call him variously an antiquarian, a mathematician, a scientist, an archaeologist, an ethnologist, a biographer, a historian, an astrologer, a botanist, a chemist, a collector, perhaps even an onomastician and a folklorist; and we would wonder what one man was doing pursuing all these interests together, and if for him they were connected or not. Today, we can understand a scientist who tries his or her hand at biography; we can just about understand a mathematician who pursues antiquarian projects when off-duty; but we would find any dabbling in astrology rather off-putting.

**Figure 1** Black lead and Indian ink miniature portrait of John Aubrey by William Faithorne the Elder, 1666. Oxford, Ashmolean Museum.

This short book, designed to accompany the exhibition on John Aubrey held by the Bodleian Library, Oxford, in the summer of 2010, explores the intellectual world of John Aubrey, and shows how Aubrey and his contemporaries divided up their intellectual pursuits, how they created new areas of research in the interstices, and how and indeed if they associated all these different ventures. In modern terms Aubrey is best thought of as a scientist, or what the seventeenth century called a 'natural philosopher', but one of the concerns of this book will be to explain how Aubrey's generation transformed traditional natural philosophy by combining previously disparate elements in ways that have themselves now stabilised as individual disciplines, some no longer recognised as belonging to 'science' in the modern way of delineating that activity. To explain this a little further, we need to understand something of how the disciplines were organised in the seventeenth century and who engaged in them.

Aubrey, like all other educated men of the early-modern period, was the product of a fundamentally different educational structure. Those who attended grammar schools concentrated largely on language study, and if they then went on to university they would spend a further seven years or so acquiring their Bachelor's and then Master's degree in arts. The Bachelor's degree in theory developed skills in the 'trivial' arts of grammar, rhetoric and logic; while the Master's degree widened this to the 'quadrivial' arts of arithmetic, geometry, astronomy and musical theory. By Aubrey's time a number of further subjects were precipitating out of these categories, notably history and geography, while pursuits such as advanced mathematics or the study of difficult languages such as Hebrew or Arabic were starting to require specialised instruction by designated professors. Something of this shift can be seen in the classifications employed by the first ever printed catalogue of a university library, namely the 1595 *Nomenclator* for Leiden University, where the 'arts' classification has been quartered into 'history', 'philosophy', 'mathematics' and 'literature'. Nevertheless, all *artistæ*, as they were known in the Latin of the academe, pursued recognisably the same kind of spread-subject degree, and their final 'arts' qualification necessarily included many things that we would not today call 'science'. Only after the acquisition of one arts degree, and usually both, could students then truly specialise, in medicine, law or theology. This system had the effect of making all *artistæ* a mixture of the humanist and the scientist, and few ever became solely one to the absolute exclusion of the other. A person who was immersed in the arts at university level was usually described as a 'philosopher', one who studied either 'moral philosophy' (ethics), 'metaphysics' or 'natural philosophy', the study of natural bodies. The natural philosopher, then, in a formulation that sounds somewhat peculiar today, was a kind of artist engaged in scientific research. Traditionally this research had been conducted mainly by reading texts, but as the seventeenth century progressed, in the academe

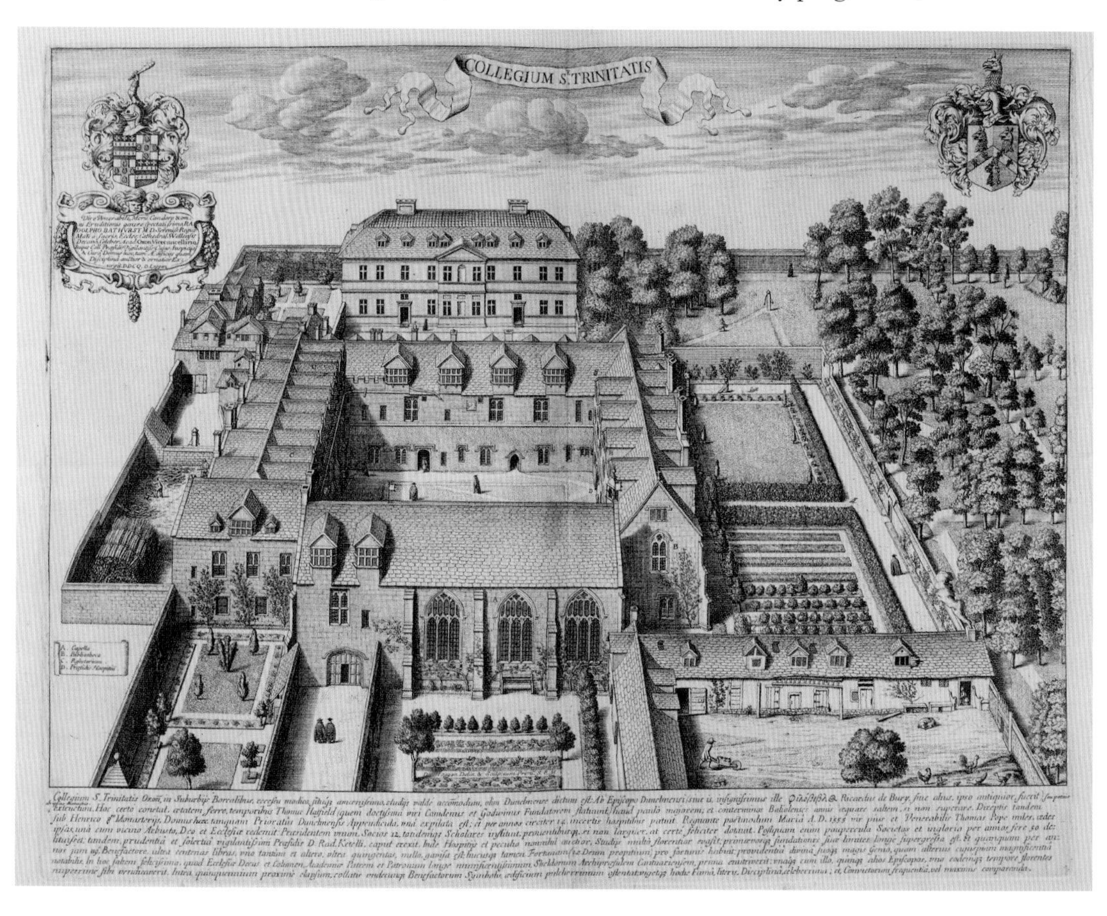


**Figure 2** Trinity College, Oxford, from John Loggan's *Oxonia Illustrata* (Oxford 1675). Oxford, Bodleian Library, Gough Maps 57, fol. 28v.

as well as in amateur contexts, reading was increasingly supplemented by and even subordinated to actual experiment, especially following the advocacy of organised and collaborative experimentation by the two great English natural philosophers William Gilbert (1544?–1603) and Francis Bacon (1561–1626).

Aubrey was educated in this kind of world. As a schoolboy he spent his time learning the Latin and Greek with which he would pepper all his later manuscripts, although he remembered fondly too that in his lonely childhood on the rural family estate he used to spend time with the tradesmen, from whom he picked up an interest in construction and mechanics. It was as a schoolboy, too, that Aubrey first came to the attention of the natural philosopher and political theorist Thomas Hobbes (1588–1679). Hobbes was little known to the world at this date, but Aubrey came to venerate the philosopher, and would later write his biography. In 1642 Aubrey went to Oxford, entering Trinity College as a 'gentleman commoner', a status reflecting his well-to-do family background. There he found the mixture of scholarship and conviviality he had missed as a child, and, although the Civil War soon shook him unwillingly from Oxford, he remained in close epistolary contact with his academic confrères, and soon made occasion to return to Trinity. It is in his letters and book purchases of this period that we first see Aubrey developing his interests in the new experimental philosophy; and it is in this period too that Aubrey started his antiquarian campaign, commissioning from a hedge priest a drawing of Osney Abbey that is our only illustration of a now demolished structure.

Aubrey, like many students, did not formally take a degree, and at any rate Royalist Aubrey's studies coincided exactly with the upheavals of the first Civil War, 1642–46. In 1646 he enrolled in the Middle Temple in London, and would often sign his books thereafter as 'Med. Templi', but there is little evidence he took law seriously. He certainly had no proficiency at it, for from the inheritance of his father's estates in 1652 until his death four and a half decades later, Aubrey's private life is a sorry narrative of ill-advised law suits, spiralling debt, bankruptcy and delitescence. But it is perhaps fortunate for us that Aubrey was quite so unfortunate himself, as his financial collapse forced him to live with and off his friends in his last decades, and it was this that kept his own social and intellectual life on the boil.

Aubrey was always a clubbable man. Trinity College for Aubrey was a club, as was the Middle Temple. In the last days of the Protectorate he joined the Rota Club, the debating forum founded by the political theorist James Harrington (1611–1677). But Aubrey took no arts degree, no law degree, and otherwise showed no interest in political theory

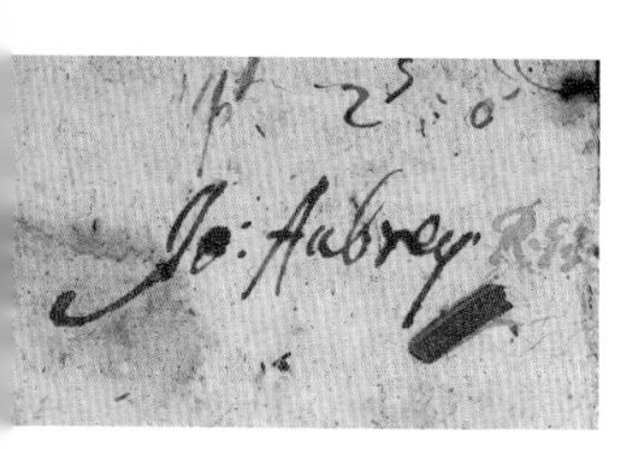

**Figure 3** Aubrey's signature on the inside of a book; following his election to the Royal Society, Aubrey habitually added 'R.S.S.' to his name. Oxford, Bodleian Library, Ashmole D 45. His formal signature in the Register of the Royal Society upon election to the fellowship is reproduced on the title page to this book. London, Royal Society, Charter Book.

per se. The one club that he was very serious about was the newly founded Royal Society of London. Aubrey was proud of his fellowship, actively attended meetings when he could, and from 1663 signs himself in his books as 'R.S.S.', 'Regalis Societatis Socius', or 'Fellow of the Royal Society'. We shall discuss the origins of the Royal Society in due course, but a few preliminary remarks may be made here about the social cohesion of this institution.

The Royal Society combined a number of different intellectual and social roles. Most importantly, it transacted experimental philosophy, and it recorded and archived these transactions. The desire to communicate its activities to a wider audience soon prompted the creation of a few new literary forms, notably the laboratory report as pioneered by Robert Boyle (1627–1691), and the scientific journal, as edited and published by Henry Oldenburg (*c.* 1619–1677), who titled his journal *Philosophical Transactions* (1665–present day). The Society also published under its own imprimatur. In addition to its own archives the Society developed a library and a repository, or a kind of working museum of natural and artificial objects. Furthermore, the Society acted as a correspondence hub, writing and receiving letters from experimentalists at home and abroad, and facilitating their interconnection. Less formally, it retained some of the ethos of a social club, generating various dining and drinking circles, and allowing its sphere of influence to extend beyond the confines of formal meetings and into the informal surrounds of coffee- and private houses. The Royal Society was not the first of its kind, and could look to previous Italian, French and English learned societies, although its direct origins lay in the interregnum experimental clubs that had sprung up in Oxford and London, with overlapping memberships, some early versions of which, as we shall see, Aubrey had encountered while a student at Trinity. Most importantly, the Society held its meetings in Gresham College, an Elizabethan educational foundation that employed professors who in theory offered public lectures, and some of whom were permanently resident in the college. In this sense the early Royal Society – a little like delitescent Aubrey in his final decades – was really the guest of another organisation. The Society in turn inspired imitative organisations, including the Oxford Philosophical Society, which met in the newly established Ashmolean Museum, the institution with which Aubrey in his final years came to identify most closely.

And yet the early Royal Society, again a little like Aubrey, was neither universally appreciated nor financially secure. The new philosophers attracted a good deal of ridicule, partly because their most widely publicised experiments struck some observers as silly – why bother weighing air?, as even Charles II was heard to muse – but partly also because for some the Society posed a threat to the religious and social order itself. Some of the philosophers asked difficult questions about matters traditionally regarded as theological property, and in any case most experimentalists seemed suspiciously keen on 'reason', traditionally the harbinger of religious unorthodoxy. Regardless of the high number of aristocrats, courtiers and churchmen among the original fellowship, such suspicions of heterodoxy persisted throughout the period, and in a few cases – including that of Aubrey – suspicion proves in retrospect well-founded. Again, unlike Louis XIV's Parisian Académie Royale des Sciences, the Royal Society achieved no state funding, and as a result had to rely on members' subscriptions. This is of considerable importance, because it means that Britain's first major scientific research institution was by definition initially composed of amateurs, men who were merely interested, but unpaid, and of necessity usually independently wealthy. There were in fact only two professional scientific employees in England in the period working outside the academe (where there was almost none either), and these were the two men employed by the Royal Society to transact its business, but who both otherwise did not

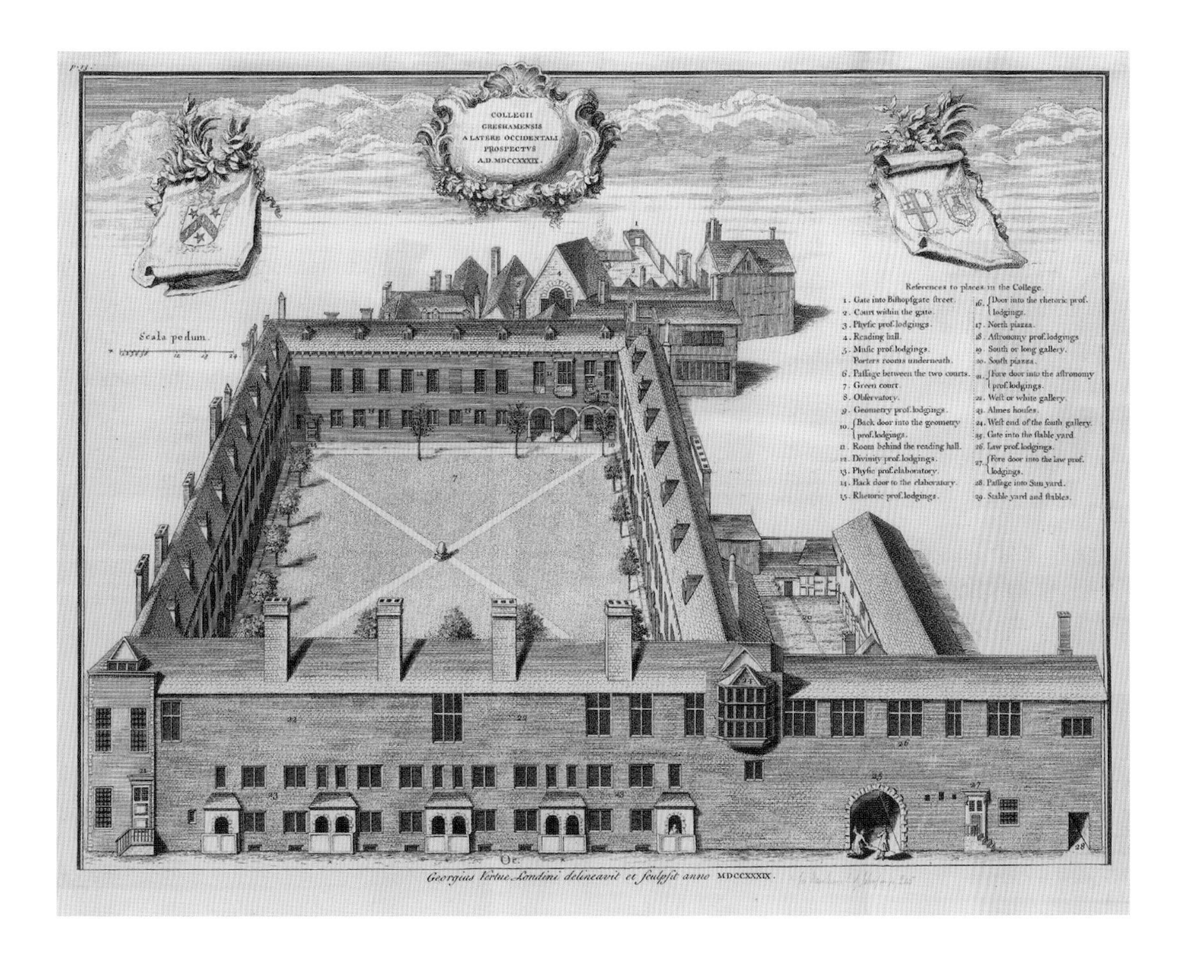


**Figure 4** Gresham College, London, by George Vertue, 1739. Oxford, Bodleian Library, Douce W. Subt. 2, plate between pp. 32 & 33.

have a separable job or much of an income. Secretary Henry Oldenburg managed the correspondence of the Society; and curator Robert Hooke (1635–1703) looked after demonstrations and books. Hooke, indeed, who was to become one of Aubrey's closest friends, may be regarded as the first professional English experimental scientist.

It is crucial to grasp this institutional angle, because it protects us from the historical Whiggism of assuming that early 'science' is to be understood merely as a preamble to later 'science' without respect to the profound metamorphosis of the institutions which generated such knowledge. But if there was no employment structure at all for science in Aubrey's period, we may be well advised to seek elsewhere for what rendered intellectual pursuit coherent. And the structure of work that most of the participants in early science shared was certainly not that of the modern research laboratory full of salaried professionals, but rather a common educational heritage that had not yet acknowledged a distinction between what we now think of as the humanities and the sciences – let alone the social sciences, also just beginning to appear. Ironically the Royal Society was itself in time to nurture this schism, and it must be conceded that its announced spectrum of interests and its peculiar way of conducting research were inherently biased towards the modern understanding of science. But this was not obvious to all or even many of the earliest fellows, and the *Philosophical Transactions* for many decades was filled with articles on inscriptions, coins, ancient pottery and the like, which in time would be classed as of (merely) historical or archaeological interest. Before 1700 in that journal, the astronomer Edmond Halley published on when precisely Julius Caesar had landed on the shores of Britain; and the experimentalist Robert Hooke conjectured about the nature of classical Chinese. The FRS and mathematician John Craige also published a separate essay in 1699 presenting the algebraic equations for the point at which Christian belief would no longer be mathematically rational. If Aubrey was an eclectic, so were most of his scholarly contemporaries.

Finally, then, what word might unite all of Aubrey's various interests? He might be called an 'experimentalist', because all of his interests involved transcending mere book study. But there is a better word, because it also carries a tang of the prejudice that the activities of the experimentalists attracted, for the non-professionals who formed the overwhelming majority of the new philosophers did acquire a name at the time, the *virtuosi*. A *virtuoso* could easily be an antiquarian, a mathematician, a scientist, an archaeologist, an ethnologist, a biographer, a historian, an astrologer, a collector – or all of the above. Aubrey, then, the non-professional, leisured and wealthy at first, harassed and impoverished at the last, was a *virtuoso* in an age in which older antiquarian pursuits were gradually intermingling with newer experimental ideas. The writer of local history was becoming interested not just in pedigrees but also in the medicinal properties of local springs; and the visitor to Stonehenge was beginning to wonder not only what it was but whether it could be geometrically analysed. Aubrey represents and is in a few cases directly responsible for this meeting of the ways, the embodiment of two passions that had been separated before his age, and that would soon separate again – the antiquarian, and the scientific. This study will use Aubrey as a lens onto those few decades in which this conjunction came to fruition.

**Figure 5** Opening of Edmond Halley's paper on when and where Caesar landed in Britain, *Philosophical Transactions* 17 (1691), pp. 495–9. Oxford, Bodleian Library, AA 74 Med, p. 495.

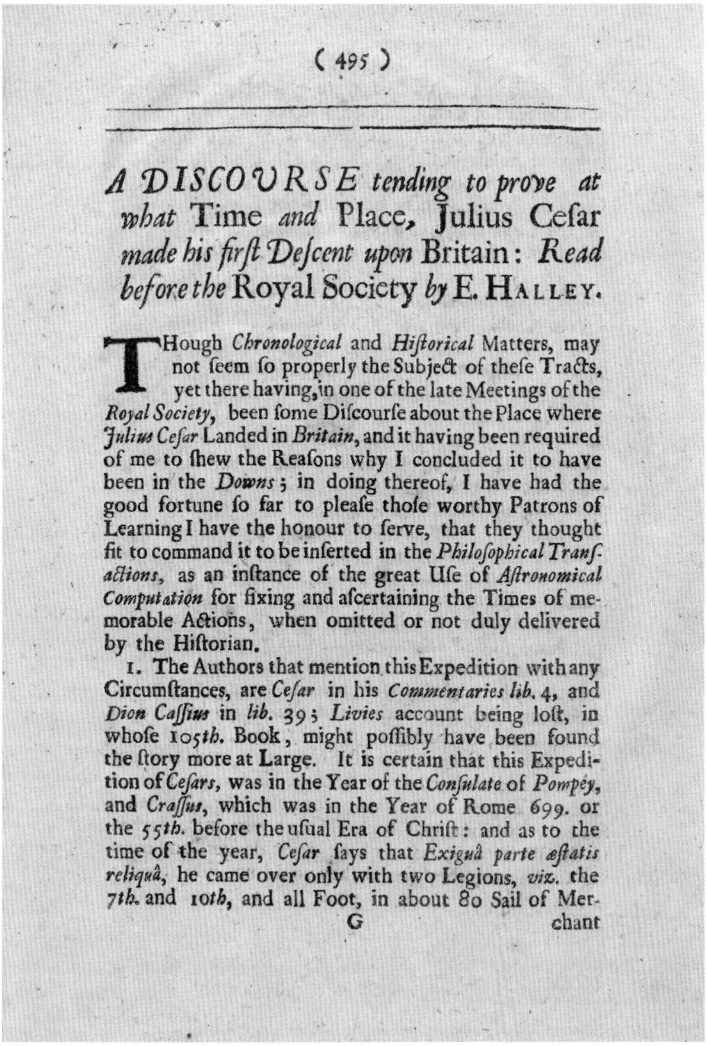

( 495 )

*A DISCOURSE tending to prove at what* Time *and* Place, Julius Ceſar *made his firſt Deſcent upon* Britain: *Read before the* Royal Society *by* E. HALLEY.

THough *Chronological* and *Hiſtorical* Matters, may not ſeem ſo properly the Subject of theſe Tracts, yet there having, in one of the late Meetings of the *Royal Society*, been ſome Diſcourſe about the Place where *Julius Ceſar* Landed in *Britain*, and it having been required of me to ſhew the Reaſons why I concluded it to have been in the *Downs*; in doing thereof, I have had the good fortune ſo far to pleaſe thoſe worthy Patrons of Learning I have the honour to ſerve, that they thought fit to command it to be inſerted in the *Philoſophical Tranſactions*, as an inſtance of the great Uſe of *Aſtronomical Computation* for fixing and aſcertaining the Times of memorable Actions, when omitted or not duly delivered by the Hiſtorian.

1. The Authors that mention this Expedition with any Circumſtances, are *Ceſar* in his *Commentaries lib.* 4, and *Dion Caſſius* in *lib.* 39; *Livies* account being loſt, in whoſe 105*th*. Book, might poſſibly have been found the ſtory more at Large. It is certain that this Expedition of *Ceſars*, was in the Year of the *Conſulate* of *Pompey*, and *Craſſus*, which was in the Year of Rome 699. or the 55*th*. before the uſual Era of Chriſt: and as to the time of the year, *Ceſar* ſays that *Exiguâ parte æſtatis reliquâ*, he came over only with two Legions, *viz*. the 7*th*. and 10*th*, and all Foot, in about 80 Sail of Mer-

G chant

# 2

# Oxford Science, London Science

The unofficial founding of the Royal Society in 1660 (its official charter dates only from 1662, renewed the following year) has often encouraged the feeling that the triumph of English science has something inherently to do with the Restoration and with London. The unspoken claim behind this assumption is that there was not much work being done before that Restoration metropolitan consolidation. Now this myth was to some extent created by the early Royal Society itself. The founding fellows were anxious to stress their devotion to the restored Crown, and Charles II was himself persuaded to accept a fellowship, duly elected FRS in January 1665. The political complexion of the founding fellows, however, was extremely varied, and fellows such as the churchman and academic John Wilkins (1614–1672), the aristocratic chemist Robert Boyle, the statistician William Petty (1623–1687) and the mathematician John Pell (1611–1685) all had strong links with the discredited Cromwellian regime. Again, the founding fellows played down their participation in the 'Hartlib Circle' of the Civil War and interregnum years, and the formerly indefatigable London intelligencer Samuel Hartlib (*c.* 1600–1662) himself finally fatigued at the Restoration, when his government pension was cancelled, and he died in poverty, his achievement unacknowledged. The man who had coordinated much of the intellectual life of the capital for decades was never even considered for the Royal Society, and is not mentioned once in Thomas Sprat's semi-official but wholly propagandist *History of the Royal-Society of London* (1667). Pre-Restoration London activity was all but written out of the acknowledged record, the Royal Society preferring instead to advertise an Oxonian genealogy, as many of the founders had collaborated there in the 1640s and 1650s. Not that this necessarily pleased the Oxonians – many of the Restoration academicians resented the new London society, seeing in it a possible competitor for their intellectual monopoly. So while the London virtuosi were reticent to acknowledge a London past, the Oxonians were reticent to grant them an Oxonian past, and otherwise opposed camps therefore colluded in obscuring the continuities in English experimentalism across the Restoration. This founding myth, which still exerts some force today, is to an extent defensible: the institutionalising of experimentalism was a crucial factor in scientific history, but it was not the only factor, nor was the Royal Society the only institution in the period. A richer way of understanding matters, as historians will always insist, is to trace continuity as well as

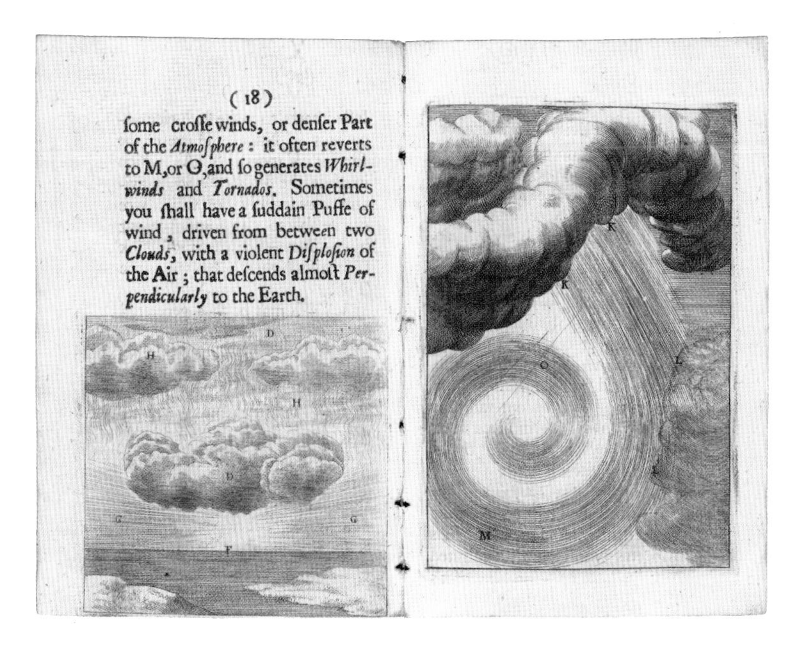

( 18 )

ſome croſſe winds, or denſer Part of the *Atmoſphere*: it often reverts to M, or O, and ſo generates *Whirlwinds* and *Tornados*. Sometimes you ſhall have a ſuddain Puffe of wind, driven from between two *Clouds*, with a violent *Diſploſion* of the Air; that deſcends almoſt *Perpendicularly* to the Earth.

**Figure 6** Ralph Bohun, *A Discourse concerning the Origine and Properties of Wind* (Oxford 1671). Oxford, Bodleian Library, Ashmole 1345, pp. 18–19.

change, and this can be done with reference to the biography of John Aubrey, who participated in all three of the Oxonian, Hartlibian and Royal Society phases.

When Aubrey arrived in Oxford in May 1642, the universities were beginning to apply themselves to the Baconian 'Great Instauration'. In Cambridge, throughout the Civil War and interregnum decades, Trinity College – Bacon's own alma mater – in particular became the academic home of soon-to-be prominent naturalists and mathematicians, notably John Ray (matriculated 1644), Isaac Barrow (1646), and Francis Willughby (1653). At Christ's College there was the future microscopist Henry Power (1641), while the important physician Francis Glisson had been at Caius for decades (1614), a college that would later attract the astronomer Laurence Rooke (1635). Nevertheless, it is Oxford that has claimed the lion's share of historical attention here, because Oxford became home to a mass of experimentalists who eventually pooled their efforts. It would be a mistake, however, to construe the rise of Oxonian 'science' in purely Baconian terms, for there had long been an independent mathematical tradition at Oxford which was to prove a decisive component in later research. Sir Henry Savile (1549–1622), for instance, was lecturing on both Ptolemy and Copernicus as early as the academic year 1570–51. Just under five decades later, in 1619, Savile founded the two chairs that still bear his name in Geometry and Astronomy, chairs that would be held in the seventeenth century by such luminaries of the early Royal Society as Christopher Wren (1632–1723) and John Wallis (1616–1703). Wallis, indeed, held his chair for a staggering fifty-four years.

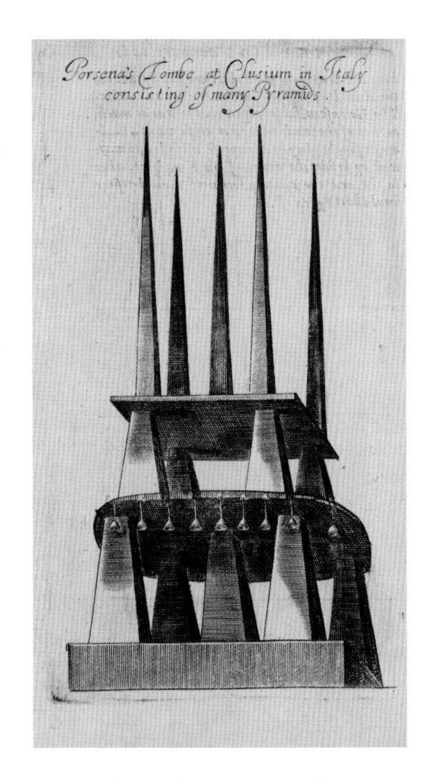


**Figure 7** The tomb of the fabled Etruscan king Lars Porsenna, as imagined in John Greaves, *Pyramidographia* (London 1646). Oxford, Bodleian Library, Savile I 7, illustration facing p. 67.

The most remarkable early incumbent of a Savilian Chair was John Greaves (1602–1652). Greaves was first a fellow of Merton, the college of Savile and of Sir Thomas Bodley, but in 1631 he joined to his Merton fellowship the London post of Gresham Professor of Geometry, the very post Robert Hooke would later hold. From 1637 to 1640 Greaves travelled in the Levant, collecting manuscripts and astronomical data, and soon after his return he succeeded to the Savilian Professorship of Geometry, as the second holder of that chair. His manuscript notebooks from the trip survive in the Bodleian, containing his pencil-drawn hieroglyphs copied down from the pyramids. His first publication, in 1646, was the *Pyramidographia, or a Description of the Pyramids in Ægypt*. Greaves had visited and measured the pyramids, taking with him various mathematical instruments from Oxford, including 'a *radius* of ten feet most accurately divided' ('into 10,000 parts', he proudly added in manuscript to a copy he presented to his brother) (1646: sg. A8r, in the Bodleian copy Savile I 7). Some of these instruments survive in Oxford today in the Museum of the History of Science, including a fine brass astrolabe originally made for Elizabeth I.

Greaves wrote as a mathematician-historian. His sensational book, about a real visit to the impossibly exotic pyramids, things of whose shape Shakespeare was famously uncertain, combined opening chapters on strictly historical themes – who built them, when, and what for – with accurate measurements of both the external faces and the internal chambers of the pyramids. Greaves showed that 'practical' mathematics was also something that could be exploited as a tool for historical enquiry. Greaves therefore also used mathematics to reconstruct ancient temples, turning textual descriptions of long-obliterated buildings into geometrical realisations. One favourite subject for such exercises was the tomb of the Etruscan king Lars Porsenna, as described in Pliny's *Natural History* from the Roman author Varro. Greaves in his study of the pyramids reconstructed this famous monument, and illustrated it in a striking plate. This would become a matter for animated discussion

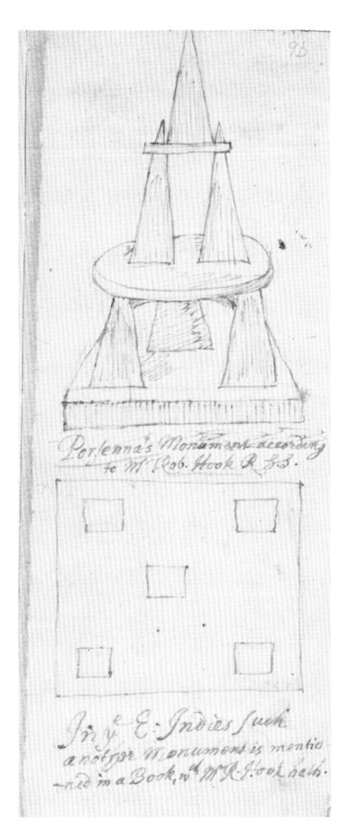


**Figure 8** Aubrey's drawing, after Hooke, of Porsenna's tomb. Oxford, Bodleian Library, MS top. gen. c. 25, fol. 9b.

later in the century by Christopher Wren, Robert Hooke, and Aubrey himself, who devoted a section of his 'Monumenta Britannica' manuscript, in the chapter on 'Barrows', to Porsenna. Aubrey too produced a drawing of the tomb, 'according to Mr Rob. Hook R.S.S.', showing both the front view and the plan of the tomb. We know Hooke owned Greaves's book, and he had evidently told Aubrey about the Porsenna illustration, but as ever Hooke thought he could do better: 'Greaves hath a draught in his... of this Monument [Aubrey obviously couldn't quite remember all the details], but Mr Hooke doeth not approve of it. Sr Christopher Wren hath made a Draught of it of about four foot long, which I could not (yet) obtain the favour to see.' Greaves drew the ancient tomb, therefore; Hooke (in his own estimation) improved on Greaves, adding a plan; and Wren the architect built a 1:75 scale model, if Varro's measurements are to be believed. It is an irony, but not one damaging the real insight of these men, that the great tomb of Lars Porsenna, King of the Etruscans, had never existed.

The Oxonian career of Greaves is a reminder of the phenomenal polymathy of the most talented seventeenth-century scholars, and Greaves, rather than Aubrey, deserves to be regarded as the first 'mathematical antiquary' in the English tradition. But we should not neglect more popular signs of a growing interest in mathematical and experimental matters. Several curricular crib manuscripts survive from the time, from which we can tell that basic training in geometry and the use of the globes came to be expected of all *artistæ*. There was a patriotic element here too: in 1600 England had seen the publication of William Gilbert's *De magnete* ('On the magnet'), perhaps the only English work of natural philosophy to find a truly international audience before the dissemination of Bacon. Gilbert argued that the earth itself was a great magnet, and turned on its axis, as did all other heavenly bodies. This furnishing of one of the core pieces of Copernicanism with a physical cause excited not just Kepler on the continent, but many lesser *artistæ* at home. In 1608 an Oxford student publicly debated the question of whether the earth was magnetic or not; three years later

**Figure 9** Domingo Gonsales's flying-machine, from Francis Godwin, *The Man in the Moone* (London 1638). Oxford, Bodleian Library, Ashmole 940(1), frontispiece.

another asked if the moon was inhabited. The Exeter fellow Nathanael Carpenter's textbook *Geographie Delineated* (1625) introduced Gilbertian ideas into the geographical curriculum. This kind of general attention eventually spawned two spectacular vernacular publications in 1638: the first piece of English science-fiction, Francis Godwin's *The Man in the Moone*; and the young John Wilkins's first publication, *The Discovery of a New World*. Both argued that the moon was a planet, and that it was possibly inhabited. Factual Wilkins was so impressed by fictional Godwin that he revised his work in 1640 to include a chapter on how to fly to the moon; Godwin had achieved the voyage by attaching his narrator to geese.

This was the Oxford to which Aubrey came, and he matriculated at the right college for such interests. A sheaf of fifteen letters preserved by Aubrey and sent to him by a friend on the fellowship there, John Lydall (1623–1657), shows Trinity to have been full of experimental activity. Only Lydall's letters to Aubrey survive, but they paint a colourful picture of experimental culture in the Trinity College of Aubrey's day. It was the college of Francis Potter (1594–1678) and Henry Gellibrand (1597–1637), both mathematicians and instrument builders. (Both men, too, constructed sundials for the college.) Potter was particularly close to Aubrey: he invented an extremely accurate beam compass that he later gave to Aubrey; and in the 1640s he experimented with blood transfusion, a priority Aubrey would later guard. The famous physician William Harvey (1578–1657), although Warden of Merton College from 1642, was also associated with the Trinity experimentalists; he and fellow physician George Bathurst kept a hen in the latter's rooms in college so that they could open eggs daily in order to study gestation. Harvey and Greaves were friends, and Greaves told the physician about the Egyptian method of hatching eggs in incubating ovens. Aubrey's letters from Lydall demonstrate ongoing interest, too, in chemical procedures, especially concerning *aurum fulminans*, or exploding gold, and in practical mathematics as it applied to astronomy and navigation.

Aubrey was soon forced by his father to leave war-torn Oxford, and after a spell at home he enrolled in 1646, as we have seen, at the Middle Temple. The admissions resister for the newly opened Sion College Library in London records his presence at this point too. A few books that Aubrey bought in the capital at this time survive, including works by Hobbes and John Selden, and Thomas Willis's 1659 *Diatribæ duæ* on fermentation, and on fevers. Thomas Willis (1621–1675), one of the foremost physicians, anatomists and chemists of his day, was already known to Aubrey from his Trinity days, as he appears in Lydall's letters as 'our Chymist'. Later, he ran a laboratory in Wadham College, which from 1649 became the hub of Oxford experimentalism, and the home of what Aubrey called the 'experimental philosophy club' (*BL* 2.301), the name under which it has been conventionally known ever since. Wilkins, now a powerful figure, had been intruded as Warden of Wadham in 1648, but unlike most other intruded figures, he proved to be successful, popular and powerful. Politically moderate, he cultivated at Wadham young scholars of varied religious and political persuasions, and he effectively turned the college into an institute for experimental research, equipped with its own collections, laboratory and botanical garden. Rules were drawn up in October 1651, setting out procedures of voting, subscription fees for purchasing equipment, grounds for expulsion, and the requirement to present experiments and to attend the weekly Thursday meeting. From Cambridge Wilkins attracted Seth Ward (1617–1689) and Lawrence Rooke (1619/20–1662); while from London came many of the members of a former London club, including Wallis, Willis, and eventually Robert Boyle. Wilkins also enlisted a few brilliant young men, notably Christopher Wren and Robert Hooke; Hooke had been working as an assistant to Willis and would subsequently be employed by Boyle. After Wilkins left Oxford to take up the mastership of Trinity College, Cambridge, in 1659, the experimental club relocated to Boyle's rooms. As Thomas Sprat, who was present, later remembered the meetings of the club,

> Their *meetings* were as frequent, as their affairs permitted: their proceedings rather by action, then discourse; chiefly attending some particular Trials, in *Chymistry*, or *Mechanicks*: they had no Rules nor Method fixd: their intention was more, to communicate to each other, their discoveries, which they could make in so narrow a compass, than an united, constant, or regular inquisition. (Sprat 1667: 56)

Aubrey would in time come to know all of these men, but he grew particularly close to Hooke. We know of around fifty men who participated in the experimental philosophy club, and many had other interconnections too: unsurprisingly, many were fellows of Wadham, with Hooke's Christ Church represented second; many had been members of the earlier London group; and many would attend the Oxford chemistry classes in the early 1660s of the Prussian chemist Peter Staehl (d. *c.* 1675), which may be considered the first classes in the subject in Oxford. Most importantly, about three-fifths of the known experimental club members would become founding fellows of the Royal Society, including Boyle, Hooke, Oldenburg, Petty, Wallis, Ward, Wilkins, Willis and Wren, to name only the better-known figures. It is to the Royal Society, and Aubrey's role in it, that we now turn.

Historians have long debated the origins of the Royal Society. Its first historian, Thomas Sprat, a Wadhamite Wilkinsian, elected FRS precisely for the purpose of writing his apologetic history of 1667, was confident that the immediate cause lay in the Oxford experimental club. John Wallis later preferred to emphasise a prior London meeting, the so-called '1645' club, which had been attended by several of the later migrants to Oxford, including Wilkins. Aubrey himself was ambiguous, in one place treating the London meetings held in William Balle's chambers in the Middle Temple as 'the first beginning of the Royal Society'; and in another place calling Wilkins's Oxford club the '*incunabula* [cradle] of the Royal Society'. If we step back from the immediate continuities between such groups, we can also observe some longer-term factors. There were continental precedents, notably the Italian Accademia dei Lincei ('Academy of the Lynxes'), to which Galileo belonged. There were also some possible English prompts. Sir Francis Kynaston (1586/7–1642), for instance, founded an academy called the 'Musaeum Minervae' in his house in Covent Garden in 1635. This institution had royal approval, and Charles I even contributed £100 to the enterprise. Although this was an educational rather than a research venture, the instruction offered is interesting. As Kynaston explained in his *The Constitutions of Musaeum Minervae* (1636), the 'Musaeum' was to be run by a 'Regent', a 'Doctor of Philosophie and Physick', and five professors. The professorships were in astronomy, geometry, music, languages and defence. The astronomy and geometry professorships were possibly influenced by Savile's two chairs, and their duties extended to practical instruction in navigation, fortification and architecture, as well as 'Analyticall Algebra' and geometry. The astronomy professor was to be Nicholas Fiske, an astrologer who described himself as an 'iatromathematicus', and who presumably therefore cast horoscopes for his medical practice. The geometer was John Speidell, a practical mathematician. The Musaeum was equipped with various instruments, and Kynaston himself appears to have been interested in magnetism, as he presented 'a curious magnetical globe' as a New Year's Gift to Charles I. Some real research work was envisaged too: Kynaston stipulated that the astronomy professor should keep a weather log and deposit it in the library, with a view towards being able to make meteorological predictions. Kynaston's Musaeum soon failed, but it is significant for three reasons: it attracted state support; it emphasised instruction in both theoretical and practical mathematics; and it encouraged research.

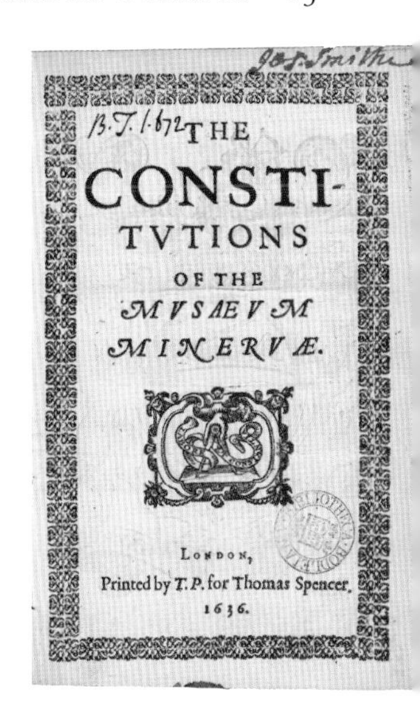
THE CONSTI-TVTIONS OF THE MVSAEVM MINERVÆ.

London,
Printed by T. P. for Thomas Spencer.
1636.

**Figure 10** Francis Kynaston, *The Constitutions of the Musæum Minervæ* (London 1636). Oxford, Bodleian Library, Gough London 216, title page.

The Musaeum Minervae also attracted the attention of the intelligencer Samuel Hartlib. Hartlib knew all about the Musaeum, and he too appreciated the research orientation possible in such an institution, looking forward to the annual printed weather logs that were promised by Kynaston (and he recorded this desire before Kynaston published his *Constitutions*, so he must have acquired this information through conversation). Hartlib himself became the foremost intelligencer of 1640s and 1650s, and recorded many more such schemes for private academies. But Hartlib himself developed a more ambitious idea. Rather than setting up reformed schools for aristocratic boys, Hartlib instead designed what he called the 'Office of Addresse', a state-authorised repository where all adults could submit and acquire information:

> Wee would advise then that a Certaine Place should be designed by the Authority of the State, whereunto all Men might freely come to give Information of the Commodities which they have to be imparted unto others; and some body should bee set in that Place to receive these Informations to the end that he may give address to every one that shall repaire to him. (Hartlib 1647: 37)

Hartlib here was inspired not simply by the plans for various academies he had heard of in the 1630s and early 1640s, but by a successful foreign example of such a bureau, the 'Bureau d'Adresse' set up by Théophraste Renaudot in Paris in 1628, again with royal support. Hartlib's bureau would disseminate 'commodities', a term which for Hartlib included intellectual and practical materials, especially ideas concerned with the advancement of technology, trade, husbandry and education.

Hartlib's plans for his state-licensed bureau ultimately came to nothing, but he ran something like it himself, collecting and swapping papers, and keeping a log of all the useful information that came his way. This log, the *Ephemerides*, constitutes the most detailed if terse record we have of intellectual exchange in this period, and it is from the *Ephemerides* too that we learn of John Aubrey's appearance in London, keen to lend assistance to the advancement of learning:

> The 2$^{d}$ of Dec[ember] [16]52 came to my house of his owne accord the first time Mr. John Aubrey dwelling at Broad Chalke neere Salisbury ... Hee seemed to be a very witty man and a mighty favourer and promoter of all Ingen[ious] and Verulamian [i.e. Baconian] designes. (*HP* 28/2/42B-43A)

Aubrey communicated to Hartlib information about Francis Potter and his many inventions, including miniature watches and threshing machines; he also told Hartlib about a possible avenue for obtaining autograph manuscripts of Bacon himself, and he commended the mathematical manuscripts of Edward Davenant. Aubrey told Hartlib about Thomas Willis and his work in the Oxford philosophical club; and Hartlib even heard in 1655 that Aubrey was writing 'the life' of Bacon, a very early indication that even in his late twenties Aubrey was known to be interested in collecting biographical information.

Aubrey led a leisurely and slightly unfixed life in the last years of the Protectorate. He travelled between his Broad Chalke estates and London. He commenced his 'Natural History of Wiltshire'. In the capital he attended James Harrington's 'Rota' club in the days of the Second Protectorate, a club that numbered many of the virtuosi in its ranks, and Aubrey was presumably attracted by the presence of the natural philosophers rather than the political discussion, as Aubrey was politically tone-deaf. Among the other attendees, however, were Robert Wood the mathematician (1621/22–1685), proponent of decimalised currency, former pupil of Oughtred and member of the Oxford experimental club; and William Petty, the great statistician and natural philosopher. Both Wood and Petty were deeply involved in Ireland at the time, and it was possibly through their influence that Aubrey himself visited Ireland in 1660 with his fellow antiquary and Middle Templer Anthony Ettrick. There Aubrey saw a manuscript in 'Saxon' script in which the Magi were described as Druids. About three months after he returned to England, the Royal Society held its first meeting.

On 28 November 1660, after Christopher Wren's usual lecture as Professor of Astronomy in Gresham College, those attending retired to the rooms of the Professor of Geometry. The dozen men present drew up a list of desirable fellows, and meetings commenced. There was a fee for membership and attendance, as well as protocols for how meetings should be conducted. Minutes were to be taken, papers were to be collected and archived, and the Society was to cultivate both a library and a repository. Meetings took place in Gresham College, but the intention was that in time a bespoke building would be purchased or constructed.

Aubrey became the 127th fellow of the Royal Society; he was proposed by the physician Walter Charleton and elected on 7 January 1663. As such, he is regarded as falling into the category of founder member. At the same meeting were also elected James, Lord Annesley, an aristocrat who proved utterly uninterested in the Society; the Wadhamite mathematician William Neile; and the Somerset clergyman and indefatigable correspondent John Beale. Charleton and Aubrey, as we shall see in a subsequent chapter, shared an interest in megaliths, and that was probably the major factor influencing Charleton's proposal of Aubrey. The two men were to remain friends until Aubrey's death: Charleton soon presented Aubrey with one of his recent books; Aubrey acquired Charleton's horoscope for his collection of genitures; and in his last surviving letter Aubrey hoped that Charleton would be the one to revise his study of folklore, the 'Remaines of Gentilisme and Judaisme' manuscript.

Aubrey was just the right kind of person for the Royal Society in its earliest decades. He was a gentleman, he was (for the time

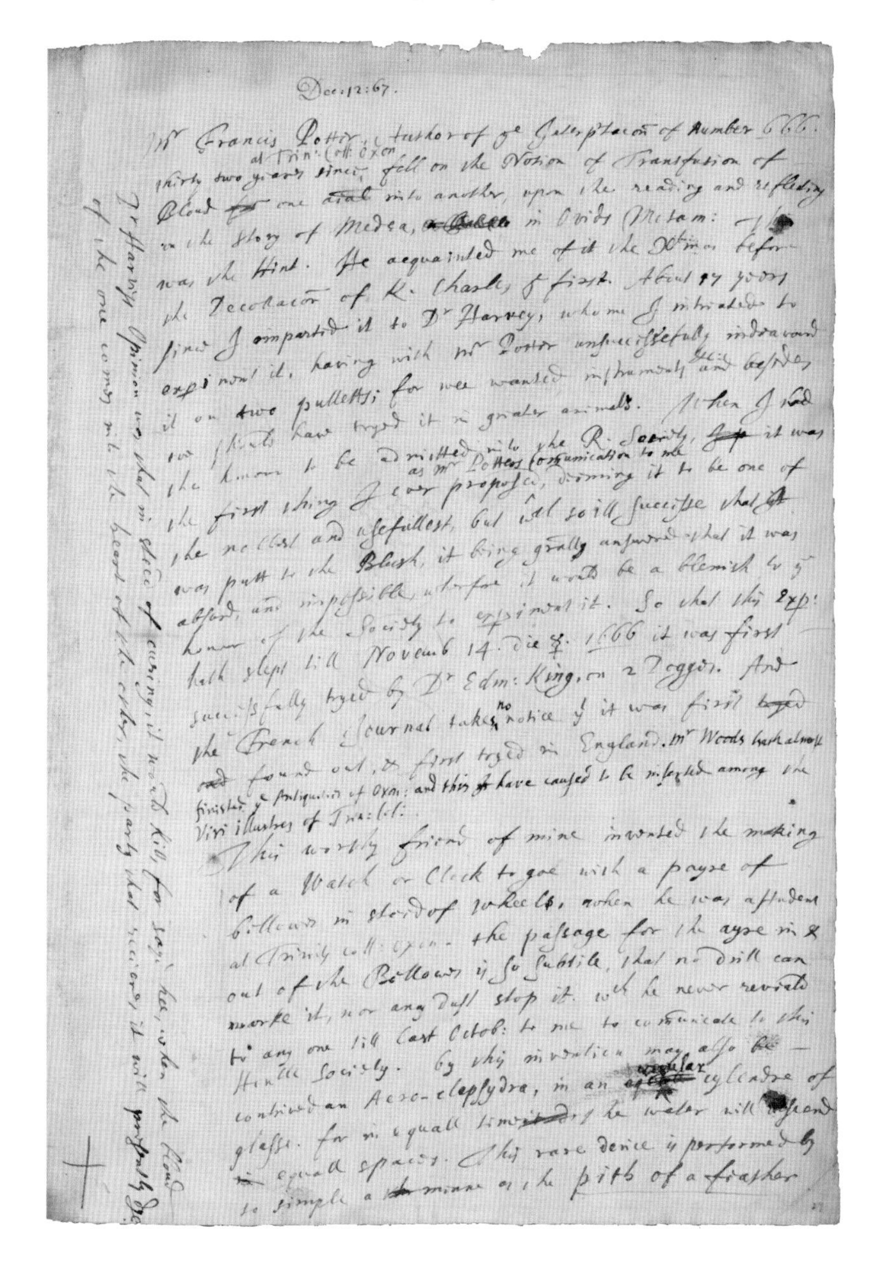

Dec: 12: 67.

Mr Francis Potter, Author of ye [illegible] of Number 666. at Trin: Coll: Oxon
thirty two yeares since, fell on the Notion of Transfusion of
Bloud from one [illegible] into another, upon the reading and reflecting
on the Story of Medea, [illegible] in Ovids Metam: This
was the Hint. He acquainted me of it the Xmas before
the Decollation of K. Charles ye first. About 17 yeares
since I imparted it to Dr Harvey, whom I introduced to [illegible]
[illegible] having with Mr Potter unsuccessfully endeavoured
experiment it, for wee wanted instruments and besides
it on two pulletts; [illegible] When I had
[illegible] have tryed it in greater animals.
the [illegible] to be admitted into the R: Society, it was
as Mr Potters communication to me [illegible] it to be one of
the first thing I ever proposed, [illegible] successe that it
the noblest and usefullest, but [illegible]
was putt to the Blush, it being greatly answered that it was
absurd, and impossible wherefore it would be a blemish to ye
honour of the Society to experiment it. So that this Exp:
hath slept till Novemb 14. die [illegible] 1666 it was first
successfully tryed by Dr Edm: King, on 2 Doggs. And
the French Journal takes no notice yt it was first tryed
and found out, & first tryed in England. Mr Woods hath almost
finished ye Antiquities of Oxon: and this I have caused to be inserted among the
Viri illustres of Trin: Coll:

This worthy friend of mine invented the making
of a Watch or Clock to goe with a payre of
Bellows in stead of wheels, when he was a student
at Trinity Coll: Oxon. the passage for the ayre in &
out of the Bellows is so subtile, that no drill can
make it, nor any Dust stop it. wch he never revealed
to any one till last Octob: to me to communicate to this
Honble Society. By this invention may also be
contrived an Aero-clepsydra, in an circular cylindre of
glasse. for in equall times the water will descend
in equall spaces. This rare device is performed by
so simple a matter as the pith of a feather.

Dr Harvey's Opinion was that in place of [illegible] it would kill, for saying [illegible] when the blood of the one comes into the other, the party that receives it will [illegible]

**Figure 11** Aubrey's letter on the first Oxford attempts to transfuse blood, claiming priority for Francis Potter. London, Royal Society, Classified Papers 12 (i), item 17, letter of 12 December 1667.

being) independently wealthy, and he knew all the right people. He was clubbable: an Oxonian, a Middle Templar and a Rotarian. His natural philosophical pedigree was sound, perhaps even a touch daunting: he recalled and championed Trinity College initiatives two decades older than the Royal Society, and pre-dating the Oxford of Wilkins. He was also extremely well-connected, and he had an antiquary's eye, as Hartlib too had found, for where caches of important manuscripts might be located. Looking at the minutes of meetings after Aubrey's election, we can see that Aubrey played to these strengths. He was very keen to tell the Royal Society all about Francis Potter's inventions, and how Potter deserved credit for priority. He presented Potter's designs for a mechanical cart without wheels, and for a clock that ran on air. Hooke was asked to investigate both (and was not convinced about the latter). More importantly, Aubrey presented Potter's work on blood transfusion in 1669, a clear attempt to remind the Society that even their most celebrated – and derided – achievements built on the work of predecessors, and that due acknowledgment was required. This sense of intellectual property was to become increasingly important to Aubrey, and we will return to the issue when we come to discuss Aubrey's biographical work.

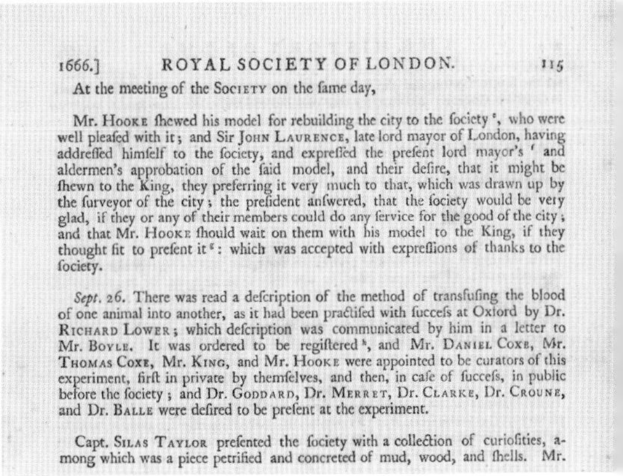

1666.] ROYAL SOCIETY OF LONDON. 115

At the meeting of the Society on the fame day,

Mr. Hooke fhewed his model for rebuilding the city to the fociety [e], who were well pleafed with it; and Sir John Laurence, late lord mayor of London, having addreffed himfelf to the fociety, and expreffed the prefent lord mayor's [f] and aldermen's approbation of the faid model, and their defire, that it might be fhewn to the King, they preferring it very much to that, which was drawn up by the furveyor of the city; the prefident anfwered, that the fociety would be very glad, if they or any of their members could do any fervice for the good of the city; and that Mr. Hooke fhould wait on them with his model to the King, if they thought fit to prefent it [g]: which was accepted with expreffions of thanks to the fociety.

*Sept.* 26. There was read a defcription of the method of transfufing the blood of one animal into another, as it had been practifed with fuccefs at Oxford by Dr. Richard Lower; which defcription was communicated by him in a letter to Mr. Boyle. It was ordered to be regiftered [h], and Mr. Daniel Coxe, Mr. Thomas Coxe, Mr. King, and Mr. Hooke were appointed to be curators of this experiment, firft in private by themfelves, and then, in cafe of fuccefs, in public before the fociety; and Dr. Goddard, Dr. Merret, Dr. Clarke, Dr. Croune, and Dr. Balle were defired to be prefent at the experiment.

Capt. Silas Taylor prefented the fociety with a collection of curiofities, among which was a piece petrified and concreted of mud, wood, and fhells. Mr.

**Figure 12** 26 September 1666: in the aftermath of the Great Fire, the Royal Society debates blood transfusion, from Thomas Birch, *The History of the Royal Society of London* (London 1756–7). Oxford, Bodleian Library, GG 33(36) Jur, vol. 2, extract from p. 115.

Aubrey made original contributions too, presenting material drawn from his county histories, as well as papers on springs and winds. He bottled spring water in Wiltshire and experimented on it in London. Aubrey seems also to have been a semi-proficient astronomer, and in 1668 he observed (and drew) a 'nubecula' or 'Clowdy Starr' between Cancer and Hydra; the drawing was accepted and archived in the Society's classified papers, where it still is. Aubrey's own interventions in meetings usually, but may not always have, won sympathy. His reports of dogs dying after licking the urine of hanged men, or of cursed shoes, for instance, will certainly have struck some of even his allies, especially Hooke, as credulous; and it is worth admitting that Aubrey was indeed regarded by some sympathetic contemporaries such as John Ray as a little too prone to superstition. (A further comment he made on how Welsh people usually have black eyes is not, however, frivolous, for Aubrey was interested in climatic determinism, and frequently tried to tie physiology and even psychology to locale.) But his interest in bringing people together – a strong Hartlibian trait – is apparent in his offers to obtain tide reports from third parties known to him, and in his presentation of letters from Andrew Paschall (1631?–1696) in Somerset on various matters, including comets and monstrous births. We will encounter Paschall again. Aubrey was frequently involved in discussions about the manuscript remains of various figures, including the medieval Oxonian pioneer Roger Bacon, as well as more recently deceased figures such as Samuel Foster, a former Gresham professor of astronomy, and the mathematician Thomas Merry, whose papers Aubrey eventually acquired from Merry's widow and presented to the Society's library. Aubrey also presented various manuscripts and books to the library, including a book in the 'Old

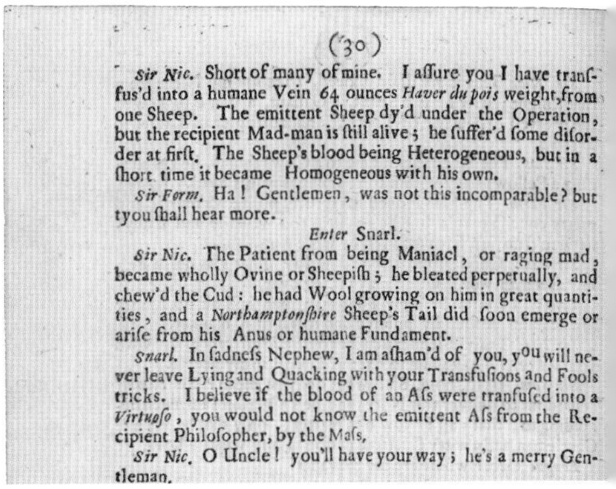

(30)

*Sir Nic.* Short of many of mine. I affure you I have tranffus'd into a humane Vein 64 ounces *Haver du pois* weight, from one Sheep. The emittent Sheep dy'd under the Operation, but the recipient Mad-man is ftill alive; he fuffer'd fome diforder at firft. The Sheep's blood being Heterogeneous, but in a fhort time it became Homogeneous with his own.

*Sir Form.* Ha! Gentlemen, was not this incomparable? but tyou fhall hear more.

*Enter* Snarl.

*Sir Nic.* The Patient from being Maniacl, or raging mad, became wholly Ovine or Sheepifh; he bleated perpetually, and chew'd the Cud: he had Wool growing on him in great quantities, and a *Northamptonfhire* Sheep's Tail did foon emerge or arife from his Anus or humane Fundament.

*Snarl.* In fadnefs Nephew, I am afham'd of you, you will never leave Lying and Quacking with your Transfufions and Fools tricks. I believe if the blood of an Afs were tranfufed into a *Virtuofo*, you would not know the emittent Afs from the Recipient Philofopher, by the Mafs,

*Sir Nic.* O Uncle! you'll have your way; he's a merry Gentleman.

**Figure 13** Transfusion experiments mocked on the stage, in Thomas Shadwell, *The Virtuoso* (London 1677). Oxford, Bodleian Library, Ashmole 1041(3), extract from p. 30.

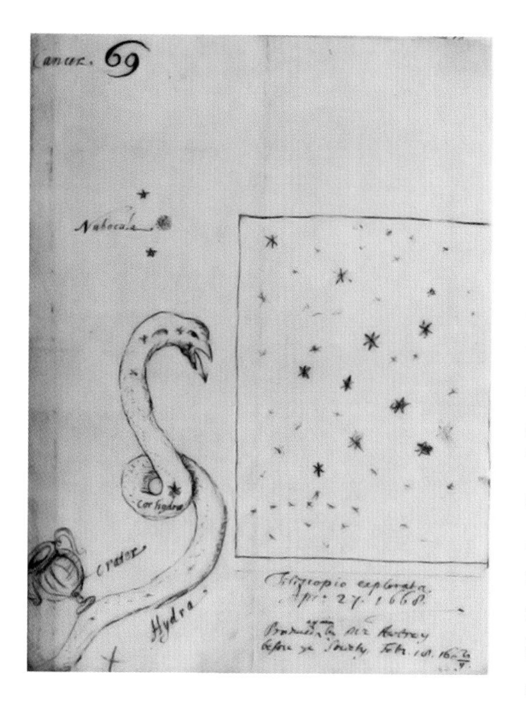

Figure 14 Aubrey at the telescope: his 'Nubecula' or 'Clowdy Starr' of 1668. London, Royal Society, Classified Papers 8 (i), item 24.

British Tongue', and Thomas Hobbes's autobiography. Again, this interest in securing texts had a part to play in priority disputes. Aubrey and Hooke, for instance, believed that the English had discovered the telescope first, before the Dutch and the Italians, and they cited – quite rightly – passages from the Elizabethan writers Leonard and Thomas Digges (father and son) to show that the Elizabethans had understood the necessary technology. Hooke read out from Thomas Digges's *Stratioticos* of 1579, and affirmed that the invention had been gleaned by Thomas's father Leonard from Roger Bacon's manuscript; Robert Plot then commented that Thomas Allen might have handed an original Roger Bacon manuscript on the topic to Leonard Digges; Aubrey then trumped them by pointing out that the account in Leonard Digges's even earlier *Pantometria* (1571) attested the same conclusion. That gave the English almost four decades on Lippershey, Metius and Galileo, and although the initial uses of the English telescope were not for astronomy, the technological precedence claimed by Aubrey and Hooke is plausible, as the passages in the Digges books (and some parallel claims by their contemporary William Bourne) do indeed describe systems of multiple lenses. The minutes make it quite clear that this was a matter of scientific patriotism – although the treasurer Abraham Hill then spoiled things a little by pointing to a 1530 passage in the Veronese physician Girolamo Fracastoro. For all their talk of 'things' over 'words', the early Royal Society fellows were exceptionally (but not inescapably) bookish.

Aubrey served the Royal Society in more mundane ways too. He audited the annual accounts on a few occasions, and in 1674 he was involved in the cataloguing of all the books, letters, accounts, instruments and gifts that had been received. Although he cannot compete for scientific stature with colleagues such as Hooke, Aubrey is in many ways the representative early FRS. He was a gentleman amateur, an exchanger of information, a donor, and an occasional administrator. He also had an intellectual stake in the two different faces of late-seventeenth-century science – the mathematical and the natural. Today we more readily recall Newton, Wren, Hooke and Halley as the stars of the period, and mathematics, whether theoretical or practical, controls the work of all four. But all four also had interests in historical and natural areas, and in statistical terms, naturalists, including chemists, probably dominated the early Royal Society, not least because an understanding of Newtonian mathematics was very hard to acquire, whereas naturalists required less training. Aubrey's friend the physician George Ent addressed a poem 'To his most honour'd Friend John Aubrey Esq.', and described Aubrey in terms that acknowledge his eclecticism, but also his grounding in natural studies:

> ... one that's studious too, whose boundlesse mind
> Scarce within Learnings compasse is confin'd;
> But chiefly Nature loves, & farre does pry
> Into her secrets with his piercing eye. (Aubrey, *Three Prose Works*, 309)

Just as the Royal Society was prompted by prior ventures, so too in turn did it prompt subsequent establishments. Aubrey's correspondent Paschall tried to create a Somersetshire network for 'a philosophical correspondence' in 1670, and in that year an observatory was opened in St Andrews University in Scotland. In Cambridge, an attempt was made to found a 'philosophick meeting' in 1684, and again in that year George Middleton and George Garden of King's College, Aberdeen, were asking for some advice from Oxford on how to set up a scientific society. All of these ventures came to little. But in Dublin, a philosophical society flourished from 1683, in close communication with the London Royal Society. Nevertheless, the most important development was back in Oxford again, where a 'meeting' was in existence by early 1683, and in May the new Ashmolean Museum was formally opened. The new Oxford club therefore met 'in ye Naturall History School, Oct: 26$^{th}$, 1683', with John Wallis in the chair. Articles were drawn up the following March, and the 'Philosophicall Society' was established, to meet every Tuesday in the Ashmolean, with a president, a secretary, a treasurer, fees for equipment costs, and even a physical format that imitated the Royal Society – only the president and secretary were allowed to sit at the central table, with the ordinary fellows seated at a distance. Aubrey was not of course a member of this society, but he did correspond with them, and several of the original members, including Robert Plot, Thomas Pigot and Ralph Bathurst, were well known to him. But the real significance here is that the Oxford Philosophical Society held its meetings not in borrowed private accommodation like the Royal Society, but in a brand new university building, the first of its kind – a public museum, lecture hall and laboratory all rolled into one. It was to this institution that Aubrey would in the end leave his papers, and not to the Royal Society. The Ashmolean was the culmination of a number of initiatives in the period to establish scientific repositories, the natural counterparts to scientific societies, and this is the subject of the next chapter.

# 3

## Repositories and Museums

Experimental science required materials on which to experiment; and in time the need for permanent collections of such materials was articulated, often both as testimonies to experiments done by and as gifts made to experimental clubs. Such collections had their origins in the antiquarian collections and the 'cabinets of curiosities' which were assembled in the sixteenth and seventeenth centuries across the continent, usually by independent scholars. The two most celebrated collections known to English scholars were the Musæum Wormianum assembled by the Danish scholar Ole Worm in Copenhagen, known through its influential printed catalogue, and the cabinet of the English Tradescants, father and son. Housed in south Lambeth, the Tradescant collection was known as the 'Ark', and could be visited by members of the public for a sixpence fee. This early museum contained both 'natural' and 'artificial' rarities, and in time it too generated a printed catalogue, the *Musæum Tradescantianum* of 1654. This collection, as we shall see, was inherited by Elias Ashmole, who augmented it and eventually gifted it to the University of Oxford, where it formed the fixed collection of the Ashmolean Museum, housed in the original Ashmolean building on Broad Street, today the Museum of the History of Science.

The need for a repository specifically suited to the purposes of active researchers and experimenters as opposed to the merely curious public had long been felt by English writers, however. Francis Bacon's fragmentary utopian fiction *New Atlantis* (1626) described the workings of 'Salomon's House', a 'college' in which experimental philosophers manipulated nature through a series of carefully planned interventions. The budding tribe of experimentalists looked to Bacon's model as a challenge, and by 1654 Walter Charleton could herald the Royal College of Physicians in London as 'Solomon's House in reality'. After the Restoration, Joseph Glanvill claimed of Solomon's House that it was a 'Prophetick Scheam of the Royal Society'.

Between Bacon's time and the creation of the Royal Society's 'Repository', however, there had been many attempts to design institutions or collections that might function as teaching or research centres. Many of these proposals were intertwined with the larger project of educational reform so central to the concerns of the Hartlib Circle. In as early as 1628 Hartlib himself had attempted to run a school in Chichester, and over the next few decades many proposals for 'reformed' schools were penned, of which the most famous today is the poet John Milton's *Of Education*. Milton's short tract was in many ways rather conservative, but the influence of the Hartlibian mentality is apparent in his insistence that his (well-born, solely male) pupils be lectured to by practical experts such as gardeners, apothecaries and mariners. Aubrey, who admired Milton if not his politics, formulated more radical proposals in his own 'Idea of Education', and as we shall later see he even provided an exact list of 'fixed' mathematical instruments to be kept in the classroom, in many cases complete with their inventors' names. These schemes for private academies, however, were soon joined by proposals for more advanced research institutions. John Evelyn (in 1659) and Abraham Cowley (in 1661) both drew up extremely detailed proposals for 'colleges', in which a permanent research staff enjoyed the provision of laboratories and repositories, and the fledgling Royal Society sought, unsuccessfully, to find a suitable property that they might buy to put such proposals into practice. But they themselves were arguably sitting in an institution that had long since realised some of these plans. Gresham College, their host institution, had in theory been providing lectures since it had become fully operational in 1597. Its seven professors, when they chose to live in, could be said to constitute a rudimentary research community. It is thanks to Aubrey's biography of Sir Kenelm Digby, too, that we know that Digby, later a great donor to the Bodleian, moved into Gresham College in 1633 for two or three years, 'where he diverted himself with his chemistry and the professors' good conversation' (*BL* 1.226–27). This was an amateur use of a professional institution, but it is significant that Digby chose to live in, rather than simply to build himself a laboratory in his own house.

Such schemes or activities took place outside the academe. But the universities were not slow to join in the discussion, and indeed to put certain proposals into action. It was well known to the English that several continental universities had been developing their own institutional collections, and Oxford in particular contributed on several fronts in the seventeenth century towards the realisation of an institution which combined the roles of a museum and a laboratory. Savile's two professors were equipped by his statutes with a library; and by 1697 the Savilian Professors' 'library' also

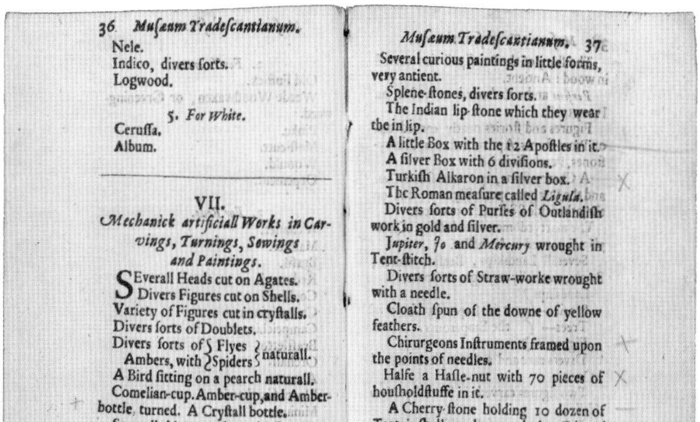

36 *Musæum Tradescantianum.*

Nele.
Indico, divers sorts.
Logwood.

5. *For White.*

Cerussa.
Album.

VII.
*Mechanick artificiall Works in Carvings, Turnings, Sowings and Paintings.*

Severall Heads cut on Agates.
Divers Figures cut on Shells.
Variety of Figures cut in crystalls.
Divers sorts of Doublets.
Divers sorts of Flyes / Ambers, with Spiders — naturall.
A Bird sitting on a pearch naturall.
Comelian-cup. Amber-cup, and Amber-bottle, turned. A Crystall bottle.
Severall things rarely cut in Corall.
Divers sorts of Corall, one with Mosse in it.
Divers things cut on Plum-stones.
Heliotropian spoone.
Many cups of Agates.
Cornelian thum-cases of the Turks.
Splene-

*Musæum Tradescantianum.* 37

Several curious paintings in little forms, very antient.
Splene-stones, divers sorts.
The Indian lip-stone which they wear in the lip.
A little Box with the 12 Apostles in it.
A silver Box with 6 divisions.
Turkish Alkaron in a silver box.
The Roman measure called *Ligula*.
Divers sorts of Purses of Outlandish work in gold and silver.
*Jupiter, Jo* and *Mercury* wrought in Tent-stitch.
Divers sorts of Straw-worke wrought with a needle.
Cloath spun of the downe of yellow feathers.
Chirurgeons Instruments framed upon the points of needles.
Halfe a Hasle-nut with 70 pieces of houshold stuffe in it.
A Cherry-stone holding 10 dozen of Tortois-shell-combs, made by *Edward Gibbons*.
A nest of 52 wooden-cups turned within each other as thin as paper.
A Hollow cut in wood, that will fit a round, square and ovall figure.

D 3 The

**Figure 15** 'Mechanick artificiall Works' from the *Musæum Tradescantianum* (London 1654). Oxford, Bodleian Library, Ashmole 1574, pp. 36–7.

contained sixteen mathematical instruments, the professors' teaching tools. John Wallis, Professor of Geometry, leased out Stable Hall at the end of New College Cloisters, and his successor Edmond Halley equipped out a turret as an observatory. Today the building is occupied by New College undergraduates, but the turret is still there, and a plaque memorialises the historic use of the building. Back across the road, the Bodleian Library itself had always held more objects that just books, perhaps the most striking of its early non-textual acquisitions being a curious alabaster mathematical pillar purchased in 1620 for £2, an obelisk symbolising various geometrical relations, perhaps a teaching tool. The Bodleian also had an ancillary function as a museum. From the beginning, it collected coins, paintings and objects of natural history as well as books, and indeed many of its oriental books remained objects to be admired rather than read until knowledge of the relevant scripts reached Oxford. In time, these collections were split between what is now Duke Humfrey's Library and the Anatomy School, and it was the job of the Assistant Librarian to conduct tours of the latter for a fee. There visitors could examine a hotchpotch of natural, anatomical and artificial rarities, ranging from fossils and skeletons to examples of miniature writing, carved hieroglyphics and a mummy, a Chinese printed book, Wingate's arithmetical tables, an entire shark, a dead pygmy, locks, whips, gloves, a witch's nipple, 'A pin vomited up by Alex. Rostorn's Daughter near Armskirk in Lancashire', and two artificial tortoises. When Thomas Hearne (1678–1735) catalogued the collection in his charge in 1705, it contained 275 exhibits and 67 coins. By 1713, the collections had grown to 415 items, and a further 130 coins. Perhaps we should take the antiquary Thomas Hearne's words with a pinch of salt when he stated that John Keill, the Savilian Professor of Geometry, declared the Anatomy School's collection 'better than that in the Laboratory', but such a claim underlines the perceived continuum in the period between the functions of the Bodleian and the more recent Ashmolean Museum.

Nevertheless, the Savilian Library and the Bodleian were still primarily repositories of texts, and the Ashmolean too would encompass two separate libraries. But there were motions afoot by the 1650s to open a new kind of research centre, 'a Magneticall, Mechanicall, and Optick Schoole, furnished with the best Instruments, and Adapted for the most useful experiments in all those faculties'. This hope was voiced in 1654 by Seth Ward, who was then Savilian Professor of Astronomy. The early 1650s had been a difficult time for Oxford, not least because of the hostility of certain radical members of the Commonwealth parliament towards the universities. In 1654 the radical preacher and physician John Webster published an intemperate attack on the universities, the

*Academiarum Examen* (1654), in which he accused the academicians of insufficient modernity, extending to their inability to do experiments. The mathematical arts, he claimed, were only partially and badly taught, and more applied subjects such as chemistry were altogether neglected. Webster styled himself as the true heir of Bacon, in opposition to the turgid and superannuated Aristotelianism which he insisted still dominated contemporary Oxford and Cambridge. Ward and Wilkins swiftly replied with their *Vindiciæ Academiarum* (1654), in which they demonstrated that most of what Webster rather shakily understood of the new science was already practised in Oxford, and in which the plans for their experimental 'Schoole' in magnetics, mechanics, and optics were also voiced.

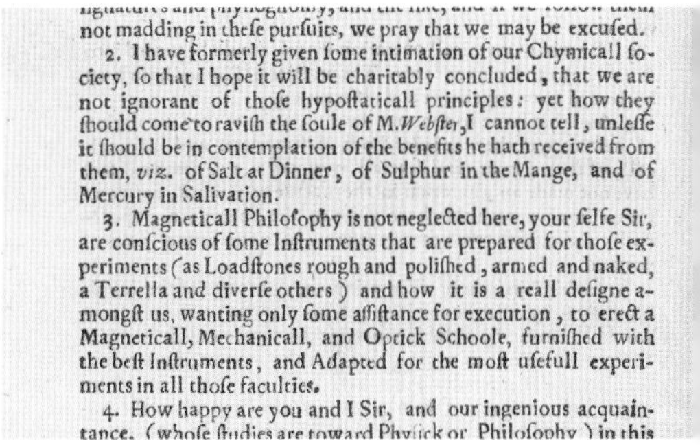

**Figure 16** Seth Ward's plans for an experimental 'Schoole' in Oxford, from his *Vindiciæ Academiarum* (Oxford 1654). Oxford, Bodleian Library, Ashmole 1621(15), extract from p. 36.

not madding in theſe purſuits, we pray that we may be excuſed.

2. I have formerly given ſome intimation of our Chymicall ſociety, ſo that I hope it will be charitably concluded, that we are not ignorant of thoſe hypoſtaticall principles: yet how they ſhould come to raviſh the ſoule of M. *Webſter*, I cannot tell, unleſſe it ſhould be in contemplation of the benefits he hath received from them, *viz.* of Salt at Dinner, of Sulphur in the Mange, and of Mercury in Salivation.

3. Magneticall Philoſophy is not neglected here, your ſelfe Sir, are conſcious of ſome Inſtruments that are prepared for thoſe experiments (as Loadſtones rough and poliſhed, armed and naked, a Terrella and diverſe others) and how it is a reall deſigne amongſt us, wanting only ſome aſſiſtance for execution, to erect a Magneticall, Mechanicall, and Optick Schoole, furniſhed with the beſt Inſtruments, and Adapted for the moſt uſefull experiments in all thoſe faculties.

4. How happy are you and I Sir, and our ingenious acquaintance, (whoſe ſtudies are toward Phyſick or Philoſophy) in this

The development of organised study, as we have seen, is associated with Wadham College, where Wilkins became Warden in 1648. Wilkins's lodgings served as a de facto meeting centre for the experimentalists, and Wilkins does appear to have developed a bespoke collection of instruments and mechanical contrivances to accompany and assist their studies, including some large magnets, a transparent beehive, and a famous 'speaking statue' in his garden, which Wilkins operated by talking into a tube that ran into his lodgings. This lively interest in mechanics had already been expressed in print by Wilkins in his popular *Mathematicall Magick* (1648), an engagingly illustrated work of 'mechanicall geometry', in which he discussed contrivances based on levers, and also the possibility of speaking automata realised by purely mechanical means. John Evelyn remembered visiting Wilkins's lodgings in mid-1654, and he has left us the fullest description of Wilkins's collection:

> We all din'd [13 July] at that most obliging & universaly Curious Dr. *Wilkins's*, at Waddum, who was the first who shew'd me the *Transparent Apiaries*, which he had built like *Castles* & *Palaces* & so ordered them one upon another, as to take the *Hony* without destroying the *Bees*; These were adorn'd with variety of *Dials*, *little Statues*, *Vanes* &c: very ornamental.... He had also contriv'd an hollow Statue which gave a Voice, & utterd words, by a long & conceald pipe which went to its mouth, whilst one spake thro it, at a good distance, & which at first was very Surprizing: He had above in his Gallery & Lodgings variety of *Shadows*, Dyals, Perspe[c]tives, places to introduce the *Species* [probably a camera obscura], & many other artif[i]cial, mathematical, Magical curiosities: a Way-Wiser [i.e. an odometer], a *Thermometer*; a monstrous *Magnes*, *Conic* & other *Sections*, a Balance on a demie Circle, most of them of his owne & that prodigious young Scholar, Mr. *Chr: Wren*, who presented me with a piece of *White Marble* he had stained with lively red very deepe, as beautifull as if it had ben naturall. (Evelyn 1955: 3.110–11)

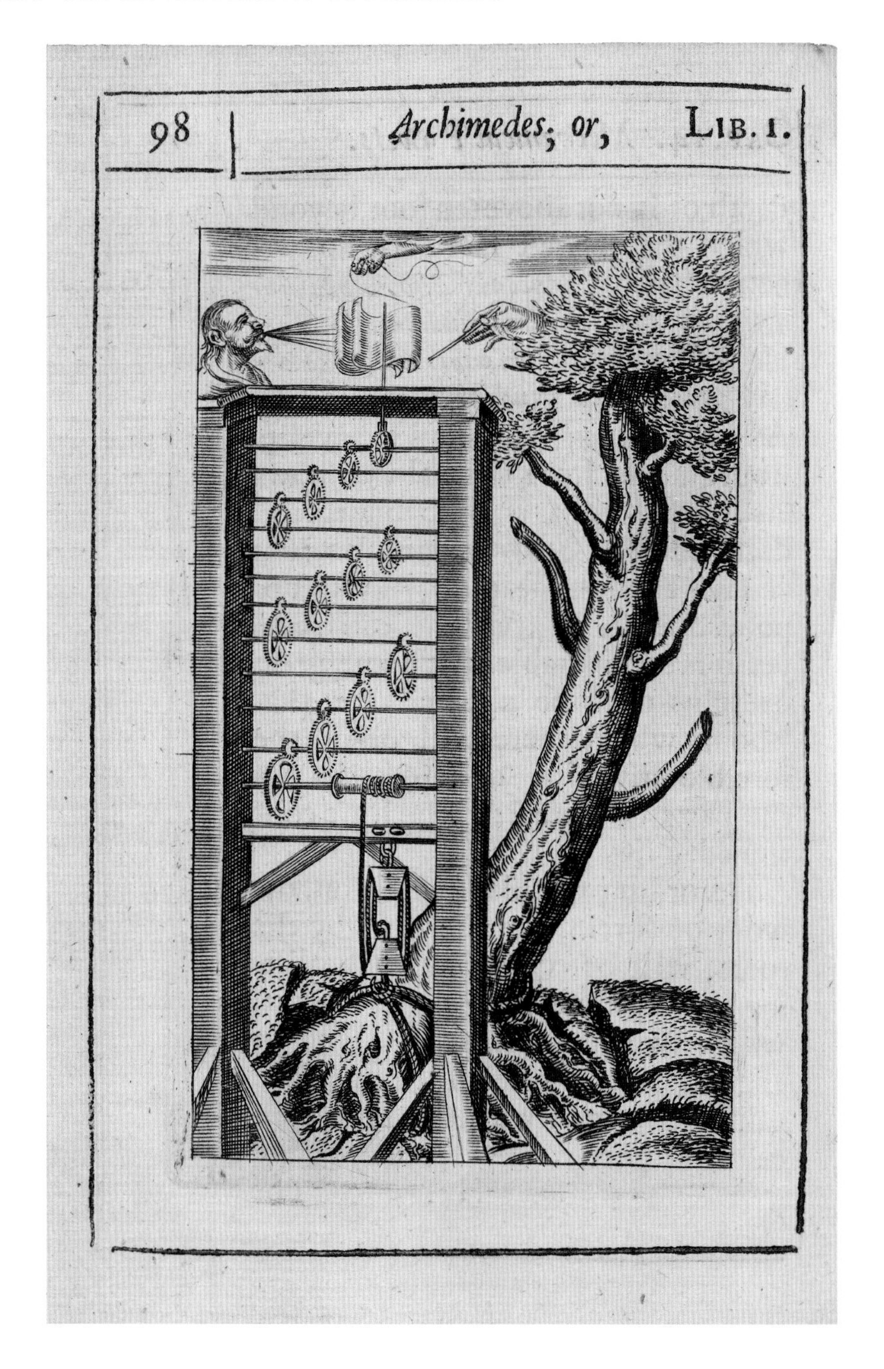


**Figure 17** The power of mechanics, from John Wilkins, *Mathematicall Magick* (London 1648). Oxford, Bodleian Library, Savile H 28, p. 98.

Wilkins's collection, in other words, consisted mainly of mechanical and mathematical instruments, but there is also evidence that Thomas Willis constructed a chemical laboratory in Wadham, and Aubrey also thought that Willis was involved with another laboratory at Peckwater Inn Chamber.

The Royal Society from its very first meetings in Gresham College resolved to use their meeting place as also their archive and their 'repository': 'Here [in Gresham] the *Royal Society* has one *publick Room* to meet in, another for a *repository* to keep their Instruments, Books, Rarities, Papers, and whatever else belongs to them' (Sprat 1667: 93). The fellows hoped in time to construct for themselves a more permanent residence, 'a *College*, which they design to build in *London*, to serve for their *Meetings*, their *Laboratories*, their *Repository*, their *Library*, and the Lodgings for their *Curators* (Sprat 1667: 434). Although the funds for this second building did not materialise, the intention, developed quickly in the first years of the Society's existence, is clear: the modern scientific society must be able to concentrate in one place all the resources needed for transacting business, combined with permanent accommodation for permanent staff. The Society, indeed, used the promise of permanence offered by corporate as opposed to private ownership of such a resource to encourage donations:

> To which Repository whatsoever is presented as rare and curious, will be with great care, together with the Donors names and their Beneficence recorded, and the things preserved for After-ages, (probably much better and safer, than in their own private Cabinets;) and in progress of Time will be employed for considerable Philosophical and Usefull purposes. (PT 1 (1666): 321)

It is likely that many items from Wilkins's Wadham collection were transferred to the Royal Society Repository, for in late 1663 it was noted that Wilkins had presented to the Society an air gun, two burning mirrors, a pair of scales, a model of a geometrical arch, a lodestone, various metal geometrical models, a concave mirror, a 'piece of perspective' (i.e. what we could call anamorphic art), as well as some petrified wood and grass, a strange bone, a coconut, and an ostrich's egg. The same minute noted that Prince Rupert had donated a water engine, Seth Ward a pendulum clock, and Boyle his famous air-pump (Birch 1756–57: 1.324). From its earliest days, therefore, the Repository encompassed both artificial and natural rarities.

The Repository itself was initially under the care of Aubrey's close friend Robert Hooke, who as Gresham Professor of Geometry already lived in his workplace, and indeed would not move from Gresham until his death. The Repository contained 'natural' materials in the form of animal, plant and mineral samples; and 'artificial' constructs, being chemical apparatus, mathematical instruments, mechanical devices, coins and pharmaceuticals. In his *History of the Royal-Society*, Sprat interestingly commented that Robert Hooke was organising the Repository to conform to the tabulation of Wilkins' philosophical language, a scheme we will discuss in a later chapter:

> This *Repository* he has begun to reduce under its several heads, according to the exact Method of the Ranks of all the *Species* of *Nature*, which has been compos'd by Doctor *Wilkins*, and will shortly be publish'd in his *Vniversal Language*. (Sprat 1667: 251)

**Figure 18** The first published account of the Royal Society's Repository and library, from Thomas Sprat, *History of the Royal-Society of London* (London 1667). Oxford, Bodleian Library, Ashmole 1649, p. 251.

*ROYAL SOCIETY.* 251

try,and practis'd here,by the care,and expence of the Duke of *Buckingham*; whom the Author of theſe Papers ought to mention with all honour; both for his Skill and Zeal in advancing ſuch *Experimental Studies* of which I am writing: and alſo becauſe it has been by the favour of ſo great a *Patron*, that I have injoy'd the leiſure, and convenience of compoſing this *Hiſtory*.

§.XXXVII *Their Repoſitory and Library.*

As ſoon as they were reduc'd into a *Fix'd Aſſembly*, one of the Principal Intentions they propos'd to accompliſh, was a General Collection of all the Effects of *Arts*,and the Common, or Monſtrous *Works* of *Nature*. This they at firſt began by the caſual *Preſents*, which either *Strangers*, or any of their own *Members* beſtow'd upon them. And in ſhort time it has increas'd ſo faſt, by a contribution from all Parts, and chiefly by the bounty of Mr. *Colwal*, that they have already drawn together into one Room, the greateſt part of all the ſeveral kinds of things, that are ſcatter'd throughout the *Univerſe*. The Keeping, and Ranging of theſe into order, is committed to Mr. *Hook*, who had alſo the honour of being made the firſt *Curator* of the *Royal Society* by election. This *Repoſitory* he has begun to reduce under its ſeveral heads, according to the exact Method of the Ranks of all the *Species* of *Nature*, which has been compos'd by Doctor *Wilkins*,and will ſhortly be publiſh'd in his *Univerſal Language*: A Work wherein this excellent Man has undertaken a Deſign, that very well fits the temper of his own Mind; for it well became him to teach a *Communion* of Speech amongſt all *Philoſophers*; whoſe chief ſtudy it has alwayes been, to promote a general agreement, and

l i 2 cor-

Aubrey himself had some stake in the Repository and its organisation. In the catalogue of the collection, published in 1681, he is recognised formally in the list of donors to the 'Musæum', having contributed 'Several Examples of MORTARS of old Castles and *Roman* Buildings … for comparing them with those now in use'; and a Roman money pot found in 1651 in Week-Field in Wiltshire, 'half full of *Roman* Coin, *Silver* and *Copper*, of several Emperors near the time of *Constantine*'. The pot appears to have been rather like a modern piggy bank, 'closed at the top, and having a Notch on one side, as in a *Christmas-Box*' (Grew 1681: 380–81). This vessel and its contents should in theory be among the collections of the British Museum, whither the contents of the Royal Society's Repository were transferred in 1781. Unfortunately all Aubrey's gifts are untraceable today. Aubrey's gifts were obviously antiquarian in inspiration, but his interest in analysing mortar is part of the new antiquarianism he championed, in which the history of technology and manufacture had a role to play. Aubrey was also a keen donor of books to the Royal Society Library; and in one case a gift by Aubrey of a medieval manuscript is revealed to be a tactical gift for swapping with Oxford. 'I have given to y$^{e}$ Library of the R. Societ[y] (amongst othe[r] good books)', Aubrey wrote to Anthony Wood in Oxford, 'a MSS of about H[enry] 4 scil[icet] Historia Roffensis. w$^{ch}$ they will exchange with you at Oxon for some Mathem. or Philosophicall book for that they will have no booke there but of those 2 sorts' (MS Wood F 39, fol. 253r).

Aubrey's association with the Royal Society's Repository goes slightly further. In 1674 the Society appears to have charged Aubrey with handling 'a Correspondence w$^{th}$ my numerous company of ingeniose Virtuosi in severall Countries' for an honorarium of £50 or £60 per annum, and in March of that year too Aubrey reported that he was busy 'writing y$^{e}$ Catalogue of y$^{e}$ Repository of ye R. S, w$^{ch}$ I doe according to y$^{t}$ incomparable Method of Dr Wilkins Philos. Gram*m*ar' (MS Wood F 39, fols. 253v, 261v). He estimated that it would take him at least a month of solid work. In other words, Aubrey, a bankrupt by this point, and, as he explains in the same letter, recently under arrest for £200 debt, was being found some paid work by his friends in the Society. In return

Aubrey undertook to correspond, and to construct a catalogue of the Repository, following in Hooke's footsteps by modelling it upon Wilkins's philosophical tables. This manuscript alas no longer survives, or has not been recognised for what it is; and when a catalogue of the Repository was finally printed, it was no longer organised upon Wilkins's model. Nehemiah Grew's *Musæum Regalis Societatis* (1681) was dedicated to the Society's treasurer Daniel Colwall (d. 1690), a wealthy merchant who had bankrolled both the purchase of the initial core of the Repository – the celebrated collection of the Londoner Robert 'Hubbard' or Hubert, which had itself generated a printed catalogue – and subsequently Grew's catalogue of it. Grew's timely catalogue appeared after the close of the difficult 1670s, a decade in which the Society ran out of momentum and into insolvency. The fellows, 'seeming to look a little pale, you [Colwall] intended hereby, to put some fresh Blood into their Cheeks', as Grew said (Grew 1681: sg. A3r–v).

Grew's catalogue was indeed riding on the crest of the Society's second major wave. The 1680s witnessed the revivification of the Royal Society, especially following the election to the fellowship in the late 1670s and early 1680s of fresh blood, including men with whom Aubrey would later have close dealings: the chemist Robert Plot, the astronomer Edmond Halley, the linguist Francis Lodwick, and the naturalist Tancred Robinson. Plot and Robinson were Oxonians, and the most

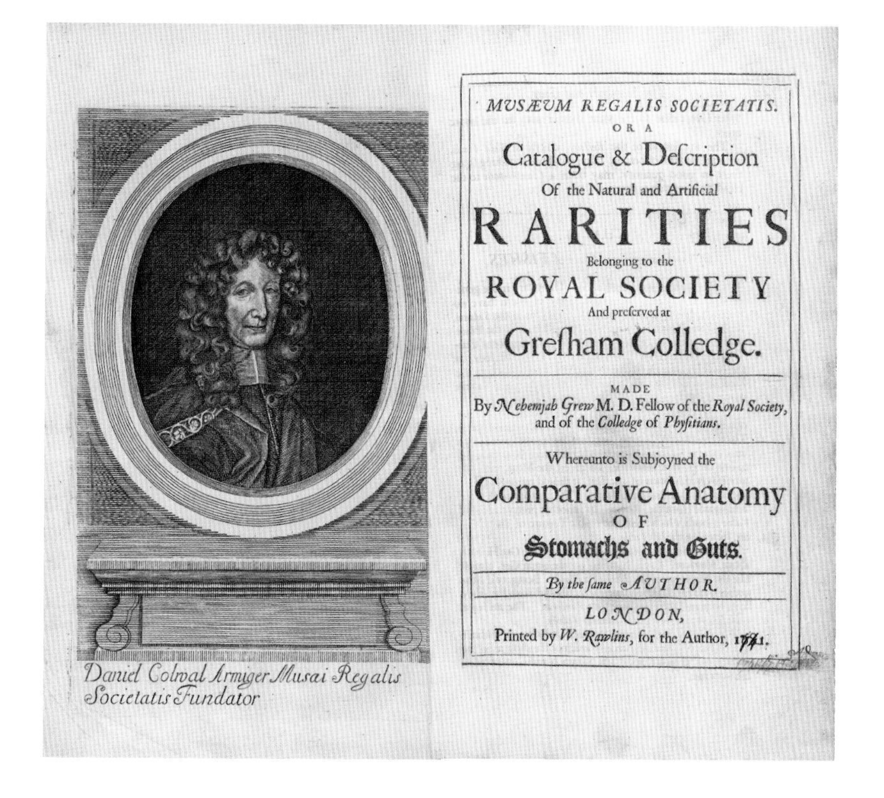


MUSÆUM REGALIS SOCIETATIS.
OR A
Catalogue & Description
Of the Natural and Artificial
RARITIES
Belonging to the
ROYAL SOCIETY
And preserved at
Gresham Colledge.

MADE
By Nehemjah Grew M. D. Fellow of the *Royal Society*, and of the *Colledge* of *Physitians*.

Whereunto is Subjoyned the
Comparative Anatomy
OF
Stomachs and Guts.

*By the same* AUTHOR.

LONDON,
Printed by *W. Rawlins*, for the Author, 1681.

**Figure 19** The first published catalogue of the Royal Society's Repository, sponsored by the merchant Daniel Colwall, and compiled by the botanist and physician Nehemiah Grew, the *Musæum Regalis Societatis* (London 1681). Oxford, Bodleian Library, Douce G subt. 3, portrait and title page.

exciting developments of the 1680s were indeed happening in Oxford. As we have seen, the Royal Society could look for its origins to several prior academic and amateur ventures and in turn it helped to spark off several new institutions. From 1679 in Oxford the Ashmolean Museum, destined to become the pinnacle of English scientific research centres in the period, had been under construction, and was formally opened in 1683. It was with this institution, and its own repository, that Aubrey would most closely identify in his final decades, gifting to it various mathematical instruments and other objects, and eventually leaving to it the majority of his own manuscript works.

The Ashmolean Museum, founded 'in imitation of the R. Society', as Ashmole himself told John Evelyn (Evelyn 1955: 4.139), possessed the unique strength of combining in one building a laboratory (basement), a lecture hall (ground floor) and the Museum proper (first floor). It had two separate libraries, one 'philosophical', the other 'chemical'. It was therefore the site simultaneously of collecting, of lecturing and of original research. The first Keeper, Robert Plot, was also the first Oxford professor of chemistry, and experimented, lectured and lived on the premises, being in effect the director of three institutions in one. Plot was also the Secretary to the Royal Society itself from 1682 to 1684. His first publication as Keeper nicely illustrates how all these functions and roles could be combined: *De origine fontium* (1685) ('On the origin of fountains'), for which the full subtitle reads 'a philosophical attempt, in a lecture held before the Philosophical Society recently established at Oxford for the purpose of promoting natural science, by Robert Plot L.L.D., custodian of the Ashmolean Museum, and Secretary of the London Royal Society.'

Aubrey knew both Ashmole and Plot very well, although he came to regard the latter as a plagiarist. Ashmole persuaded Aubrey to gift his manuscripts and some objects to the Museum, and Aubrey consented, perceiving in the Ashmolean a security that he could not guarantee from his collaborator the cantankerous Oxford antiquary Anthony Wood (1632–1695), whom Aubrey had some trouble persuading to hand over the manuscripts in his care to the museum. Aubrey's name was eventually entered into

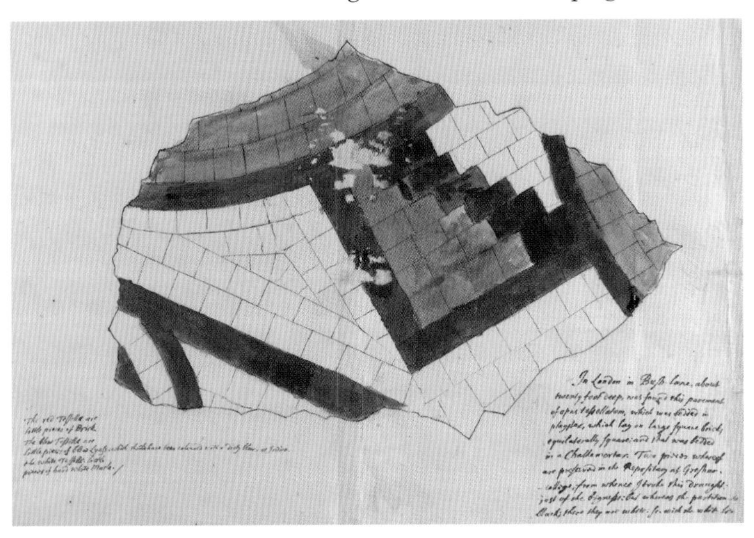

**Figure 20** Aubrey's drawing of a Roman pavement found 20 foot under Bush Lane in London, and deposited in the Royal Society's Repository. Oxford, Bodleian Library, MS top. gen. c 25, fol. 108r.

the benefactors' book of the Ashmolean in 1679, where about eighty books and manuscripts, various pictures, Roman coins and other objects are mentioned. He donated a portrait of himself, too, in 1688, which was alas then stolen a few years later. He acquired and gifted part of a Roman mosaic pavement, and drew a colour illustration of it, now part of his 'Monumenta Britannica' collections. Perhaps the most peculiar donation, however, was a pair of plaster casts of the hands of Judith Dobson, wife to the celebrated portraitist William Dobson (1611–1646), an example of Aubrey's interest in technologies for artistic production, although Aubrey will have savoured, too, the specific provenance. According to Aubrey, it was the news that he was transferring all his papers to the Ashmolean that prompted the Royal Society to commission a copy of his 'Natural History of Wiltshire'. Nevertheless, Aubrey remained committed to the new repository, not least because of a final, firm friendship he struck up with an Oxonian. In 1687 the Welsh polymath Edward Lhuyd (1659/60?–1709) was appointed Plot's assistant; Lhuyd had been a member of the Philosophical Society, and had worked as a chemical demonstrator for Plot's lectures since 1684. Upon Plot's retirement in 1691, Lhuyd was appointed Keeper in his place, and it was with Lhuyd that Aubrey corresponded most frequently in his final years; around forty letters between the two men survive. Lhuyd catalogued Aubrey's books and manuscripts, had them bound, and encouraged Aubrey to keep donating to the Ashmolean. It is thanks to Lhuyd, therefore, that we possess Aubrey's extraordinary letter books, two huge manuscripts of incoming autograph letters arranged by surname (MSS Aubrey 12, 13). Aubrey was fortunate in Lhuyd: the younger scholar shared all of Aubrey's interests, and treated Aubrey with much more respect than Plot had done. That almost all of Aubrey's manuscripts have survived safely is thanks largely to the establishment of the Ashmolean at just the right time, and to the diligence of its second Keeper.

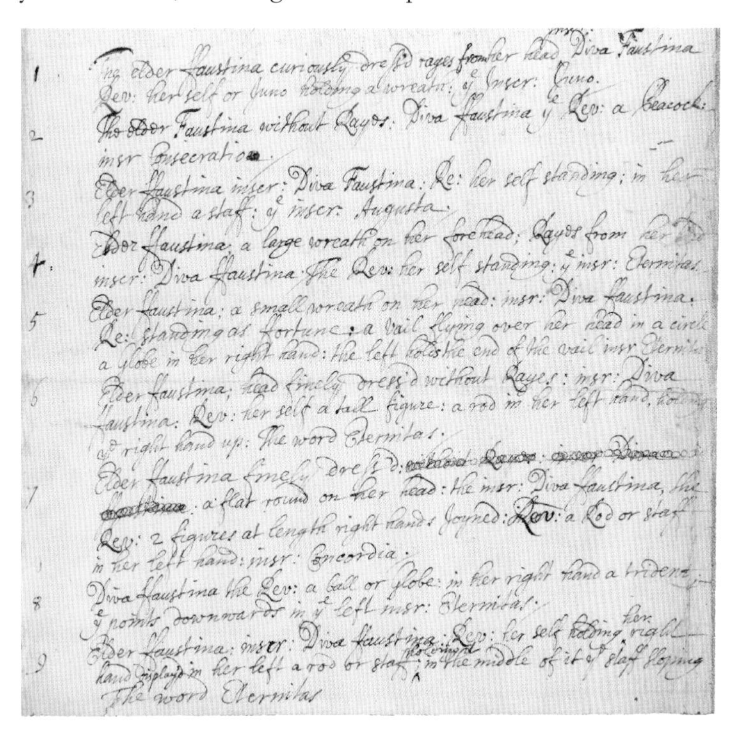

1 The elder Faustina curiously dress'd rayes from her head Diva Faustina
Rev: her self or Juno holding a wreath: ye Inscr: Juno.
2 The elder Faustina without Rayes: Diva Faustina ye Rev: a Peacock:
insr Consecratio.
3 Elder Faustina inscr: Diva Faustina: Re: her self standing: in her
left hand a staff: ye inscr: Augusta.
4 Elder Faustina a large wreath on her forehead; Rayes from her
inscr: Diva Faustina The Rev: her self standing: ye insr: Eternitas.
5 Elder Faustina; a small wreath on her head: insr: Diva Faustina.
Re: standing as Fortune; a vail flying over her head in a circle
a globe in her right hand: the left holds the end of the vail insr Eternitas
6 Elder Faustina; head finely dress'd without Rayes: insr: Diva
Faustina: Rev: her self a tall figure: a rod in her left hand, holding
ye right hand up: The word Eternitas.
7 Elder Faustina finely dress'd: a flat round on her head: the insr: Diva Faustina, the
Rev: 2 figures at length right hands Joyned: Rev: a Rod or staff
in her left hand: insr: Concordia.
8 Diva Faustina the Rev: a ball or globe: in her right hand a trident
ye points downwards in ye left insr: Eternitas:
9 Elder Faustina: insr: Diva Faustina: Rev: her self holding her right
hand display'd in her left a rod or staff holding in the middle of it ye staff sloping
The word Eternitas

**Figure 21** The Wanborough coin hoard of 1688: Roman Imperial coins concealed in the reign of the Emperor Commodus and acquired by Aubrey's friend Sir James Long. Oxford, Bodleian Library, MS top. gen. c 25, extract from fol. 115a.

# 4

# Aubrey and Mathematics

The common versions of Aubrey are Aubrey the antiquary and Aubrey the biographer. These versions are true, but incomplete. For, perhaps somewhat surprisingly, Aubrey was deeply influenced by a discipline seemingly irrelevant to those pursuits, namely mathematics. Aubrey's commitment to mathematics shaped his attitude to the rest of his intellectual interests perhaps more than any other single commitment, and accordingly we turn to a discussion of mathematics as it was encountered by Aubrey, and what he in turn drew from it.

Mathematics was once regarded by historians as a neglected subject in medieval and early-modern education, suddenly transformed by a few men in the seventeenth century working largely outside of the confines of the universities. This view has proved hard to sustain. Many of the great names of English mathematics – John Wallis, Isaac Barrow, Isaac Newton, for instance – worked within the academe; and while it is true that the very successes in mathematics in the period also gradually elevated the subject above the reach of the general student, it was precisely the growing respect for mathematical training in the early-modern university that laid the platform for the advances that would eventually render advanced mathematics a specialism for the few. Geometry was increasingly likened to logic, hitherto regarded as the undisputed ruler of the *artes*. Aubrey recalled how Thomas Hobbes received his philosophical conversion not from Aristotle, whom he despised, but from Euclid:

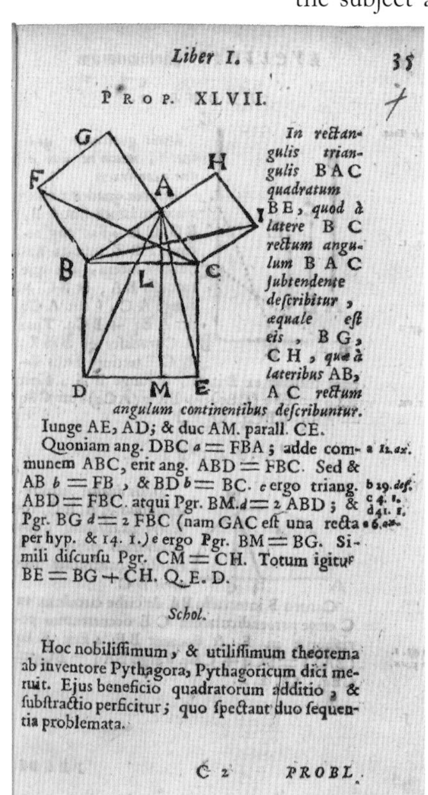

*Liber I.* 35

PROP. XLVII.

*In rectangulis triangulis* BAC *quadratum* BE, *quod à latere* BC *rectum angulum* BAC *ſubtendente deſcribitur, æquale eſt eis,* BG, CH, *quæ à lateribus* AB, AC *rectum angulum continentibus deſcribuntur.*

Iunge AE, AD; & duc AM. parall. CE.

Quoniam ang. DBC *a* = FBA; adde communem ABC, erit ang. ABD = FBC. Sed & AB *b* = FB, & BD *b* = BC. *c* ergo triang. ABD = FBC. atqui Pgr. BM. *d* = 2 ABD; & Pgr. BG *d* = 2 FBC (nam GAC eſt una recta *e* per hyp. & 14. 1.) *e* ergo Pgr. BM = BG. Simili diſcurſu Pgr. CM = CH. Totum igitur BE = BG + CH. Q.E.D.

a 12. ax. b 29. def. c 4. 1. d 41. 1. e 6. ax.

*Schol.*

Hoc nobiliſſimum, & utiliſſimum theorema ab inventore Pythagora, Pythagoricum dici meruit. Ejus beneficio quadratorum additio, & ſubſtractio perficitur; quo ſpectant duo ſequentia problemata.

C 2 PROBL.

**Figure 22** Hobbes discovers geometry: the 'Pythagoras' proposition from Isaac Barrow's edition of Euclid (London 1659). Oxford, Bodleian Library, Lister I 107, p. 35.

> He was 40 yeares old before he looked on Geometry; which happened accidentally. Being in a Gentleman's Library, Euclid's Elements lay open, and 'twas the *47 El. libri I.* He read the Proposition. *By G–*, sayd he, (he would now and then sweare an emphaticall Oath by way of emphasis) *this is impossible!* So he reads the Demonstration of it, which referred him back to such a Proposition; which proposition he also read. *Et sic deinceps* [and so on] that at last he was demonstratively convinced of that trueth. This made him in love with Geometry.

The proposition Hobbes lit upon happens, a little too conveniently, to be what we know as 'Pythagoras' theorem', and so perhaps Aubrey's account has to be read with some caution. But Hobbes's respect for geometrical method is not in doubt, and he later claimed famously in *Leviathan* (1651) that geometry was 'the only science that it hath pleased God hitherto to bestow on mankind'. In 1657, the 24-year-old Christopher Wren took up the Gresham Chair of Astronomy, and in his inaugural speech he too enthroned mathematics:

> Mathematical Demonstration being built upon the impregnable Foundations of Geometry and Arithmetick, are the only Truths, that can sink into the Mind of Man, void of all Uncertainty; and all other Discourses participate more or less of Truth, according as their Subjects are more or less capable of Mathematical Demonstrations. (Wren 1750: 200–201)

According to Aubrey, Hobbes's discovery of Euclid took place in the late 1620s, long before Aubrey went to Oxford. But the mathematical arts were already widely taught in the universities by that point, and in Oxford various colleges had appointed designated mathematical lecturers by that date. Most significantly, Sir Henry Savile founded the two Savilian Professorships of Geometry and Astronomy in 1619, and increasingly these professors, who lectured in what are now the lower reading rooms of the Bodleian, took over the more complex aspects of geometry and astronomy, leaving more basic instruction to the colleges. The two professors also maintained a mathematical library in the study connecting the two wings of the Old Schools Quadrangle. For personal study, students had an increasing array of textbooks they might acquire in all areas of mathematics. Algebra, for instance, was often learnt from William Oughtred's *Clavis mathematicæ*, which was in its fourth edition by 1667. Oughtred had a particular importance for Aubrey, as he was also a keen chemist and astrologer, and he taught a whole generation of mathematical giants, including John Wallis, Seth Ward, and Christopher Wren. Aubrey described how Oughtred 'would drawe lines and diagrams on the Dust' (MS Aubrey 6, fol. 39r), surely an anecdote (or a practice) in imitation of Archimedes, who drew lines obsessively. Aubrey later sought out and purchased at least one of Oughtred's own annotated copies of mathematical texts, as we shall see, and he acquired an autograph letter from Oughtred to Seth Ward concerning the publication of the third edition of his *Clavis* (1652), which Aubrey stitched into his life of the mathematician. Robert Hooke's own copy of this third edition of Oughtred also survives, marked up by the young experimentalist (British Library 529 b. 19 (4–5)). Later, John Wallis's *A Treatise of Algebra* (1685) was the first published English history of that subject, and emphasised, as Aubrey would in his biographical work, the contributions of English mathematicians.

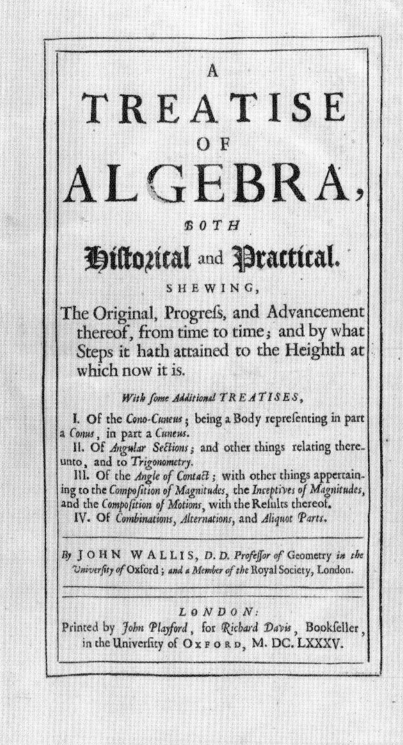

A

TREATISE

OF

ALGEBRA,

BOTH

Hiſtorical and Practical.

SHEWING,

The Original, Progreſs, and Advancement thereof, from time to time; and by what Steps it hath attained to the Heighth at which now it is.

*With ſome Additional* TREATISES,

I. Of the *Cono-Cuneus*; being a Body repreſenting in part a *Conus*, in part a *Cuneus*.
II. Of *Angular Sections*; and other things relating thereunto, and to *Trigonometry*.
III. Of the *Angle of Contact*; with other things appertaining to the *Compoſition of Magnitudes*, the *Inceptives of Magnitudes*, and the *Compoſition of Motions*, with the Reſults thereof.
IV. Of *Combinations*, *Alternations*, and *Aliquot Parts*.

*By* JOHN WALLIS, *D. D. Profeſſor of* Geometry *in the Univerſity of* Oxford; *and a Member of the* Royal Society, London.

*LONDON:*
Printed by *John Playford*, for *Richard Davis*, Bookſeller, in the Univerſity of OXFORD, M. DC. LXXXV.

**Figure 23** The first English-language history of algebra: John Wallis's *A Treatise of Algebra* (London 1685). Oxford, Bodleian Library, Savile A 3(2), title page.

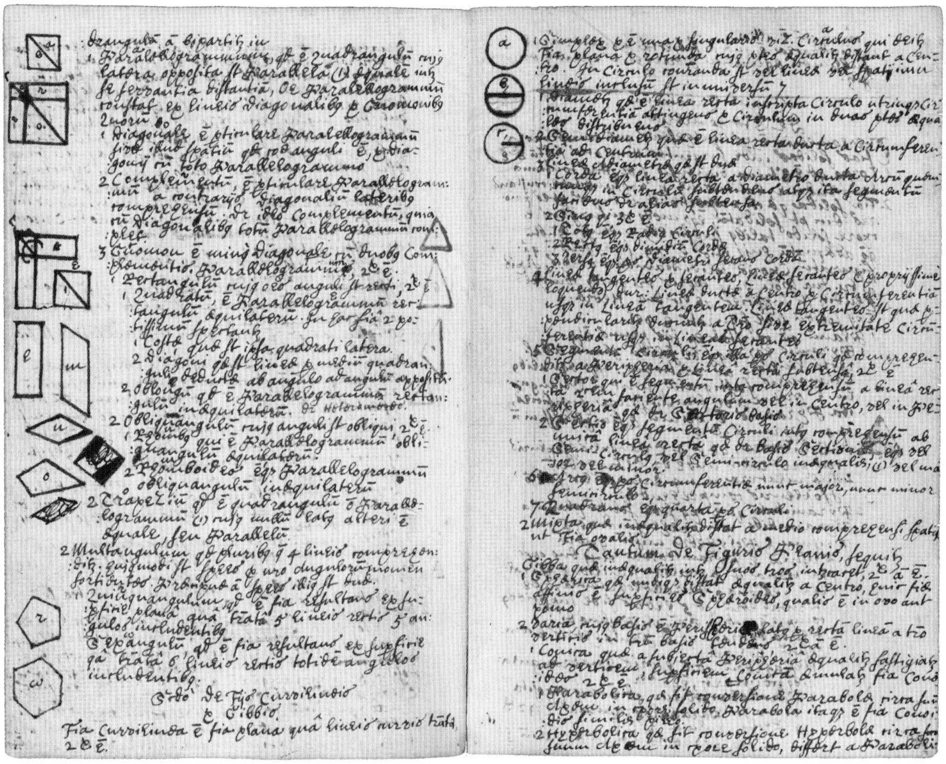

**Figure 24** Undergraduate cribs: a set of *systemata*, including one on geometry, acquired by the future Regius Professor of Greek Humphrey Hody as a fresher for 2s. on 13 March 1676, three days after he matriculated. Oxford, Bodleian Library, MS Rawlinson D 1121, fols 110v–111r.

*Top right* **Figure 25** Aubrey's own Euclid, in the London 1620 edition of Commandinus. He acquired it in 1648, while a student at Trinity College. Oxford, Worcester College, title page.

*Bottom right* **Figure 26** One of Aubrey's mathematical gifts to Gloucester Hall: Robert Recorde, *The Whetstone of Witte* (London 1557). Oxford, Worcester College, title page.

Mathematics was coming to be seen, therefore, as a desirable subject, even as the supremely rational activity. As Francis Osborne wrote in his Oxonian best-seller *Advice to a Son* (1st edn, 1655), also owned by Aubrey, '*Mathematicks* [is] the Queen of Truth, that imposeth nothing upon her Subjects, but what she proves due to belief by infallible demonstration' (Osborne 1656). Yet it should not be thought that mathematics was taught solely as a theoretical discipline. Most college libraries owned terrestrial and celestial globes, and all *artistæ* were expected to be able to manipulate both, and to acquire basic geometrical proficiency. The Savilian professors, too, quickly built up a store of more complicated mathematical instruments. A few instrument makers set up shop in Oxford, usually in the vicinity of Holywell, near the traditional stalls and shops of the Oxford stationers, where students would buy their first- and second-hand books. Many students owned their own mathematical instruments, including sectors, slide rules and quadrants. More still probably owned paper equivalents, most of which have of course perished. Aubrey himself was very keen on instrumentation, and would donate several instruments to the Ashmolean.

We can glean details of Aubrey's own experiences learning mathematics from his letters from John Lydall of Trinity between 1649 and 1651, in which the practical applications of mathematics bulk large.

Lydall obviously taught a little mathematics for his college, but he still asked – and obtained – advice from Aubrey about a book on dialling, or the construction of sundials. Other books they discussed were on magnetism, geography, navigation, trigonometry and the use of the sector. These are all notably 'applied' areas, and it would seem, therefore, that many men in the Trinity College of Aubrey's time appreciated the instrumental aspect of mathematics, just as they enjoyed chemical experimentation. Several of the books Aubrey signed in this period as having been purchased by him while a student at Trinity College have survived, chiefly concerning astronomy and arithmetic: William Blaeu's *Institutio Astronomica* (1640), Federicus Commandinus' *Astronomica* (1620), Michael Maestil's *Epitome Astronomiæ* (1624) and Robert Recorde's *Arithmeticke* (1615). No doubt many such books have been lost, but Aubrey's later comment that Edmund Gunter's '*Booke of the Quadrant, Sector and Crosse-Staffe* … made young men in love with that Studie' (*Lives* 1.276), and the survival of Aubrey's own copy of this work, suggest that it too derives from this period. Later purchasing inscriptions by Aubrey show that he continued to buy, in ever greater numbers, mathematical books until the end of his life, although many he gave away, or was eventually forced to sell.

Jo: Aubrey Trin: Coll: Oxon. 1648.

ΕΥΚΛΕΙΔΟΥ
ΣΤΟΙΧΕΙΩΝ
ΒΙΒΛΙΑ ιγ.
ELEMENTORVM
EVCLIDIS LIBRI
TREDECIM.
SECVNDVM VETERA
exemplaria restituti.
EX VERSIONE LATINA
FEDERICI COMMANDINI
aliquam multis in locis
castigata.

LONDINI,
Excudebat Gulielmus Iones.
M.DC.XX.

Aubrey continued to enjoy doing mathematics throughout his adult years, remarking that he even solved problems on horseback or in the lavatory (*BL* 1.42). He made no original contributions to mathematical theory himself, but as ever he encouraged others. As he felt his earliest instruction in mathematics had been unsound, he received further help from men such as Edward Davenant, vicar of Gillingham in Wiltshire, and nephew of Bishop Davenant of Salisbury. Despite Aubrey's encouragement, Edward Davenant would not publish his works, however, preferring to release them as small manuscript booklets, and among Aubrey's mathematical manuscripts in Worcester College, Oxford, there is a copy of Davenant's 'Algebra Literalis' manuscript, bearing Aubrey's annotation, dated 1659, 'This Algebra I transcribed from y$^{e}$ MS of M$^{ris}$ Anne Ettrick the eldest daughter of D$^{r}$ Davenant: who is a very good Logist.' (It is uncertain who is the 'very good Logist' (i.e. versed in logarithms), but it seems plausible that it is Anne herself, who married one of Aubrey's best friends, Anthony Ettrick; we know Aubrey was interested in mathematical women.) This particular tract is further notable, because much later Aubrey commissioned what he called 'Daventismi Pelliati' (i.e. 'Davenant's materials

The whetstone of witte,
whiche is the seconde parte of Arithmetike: containyng thextraction of Rootes: The Cossike practise, with the rule of Equation: and the woorkes of Surde Nombers.

Though many stones doe beare greate price,
The whetstone is for exersice
As neadefull, and in woorke as straunge:
Dulle thinges and harde it will so chaunge,
And make them sharpe, to right good vse:
All artesmen knowe, thei can not chuse,
But vse his helpe: yet as men see,
Noe sharpenesse semeth in it to bee.
The grounde of artes did brede this stone:
His vse is greate, and moare then one.
Here if you list your wittes to whette,
Moche sharpenesse therby shall you gette.
Dulle wittes hereby doe greately mende,
Sharpe wittes are fined to their fulle ende.
Now proue, and praise, as you doe finde,
And to your self be not vnkinde.

These Bookes are to bee solde, at the Weste doore of Poules, by Jhon Kyngstone.

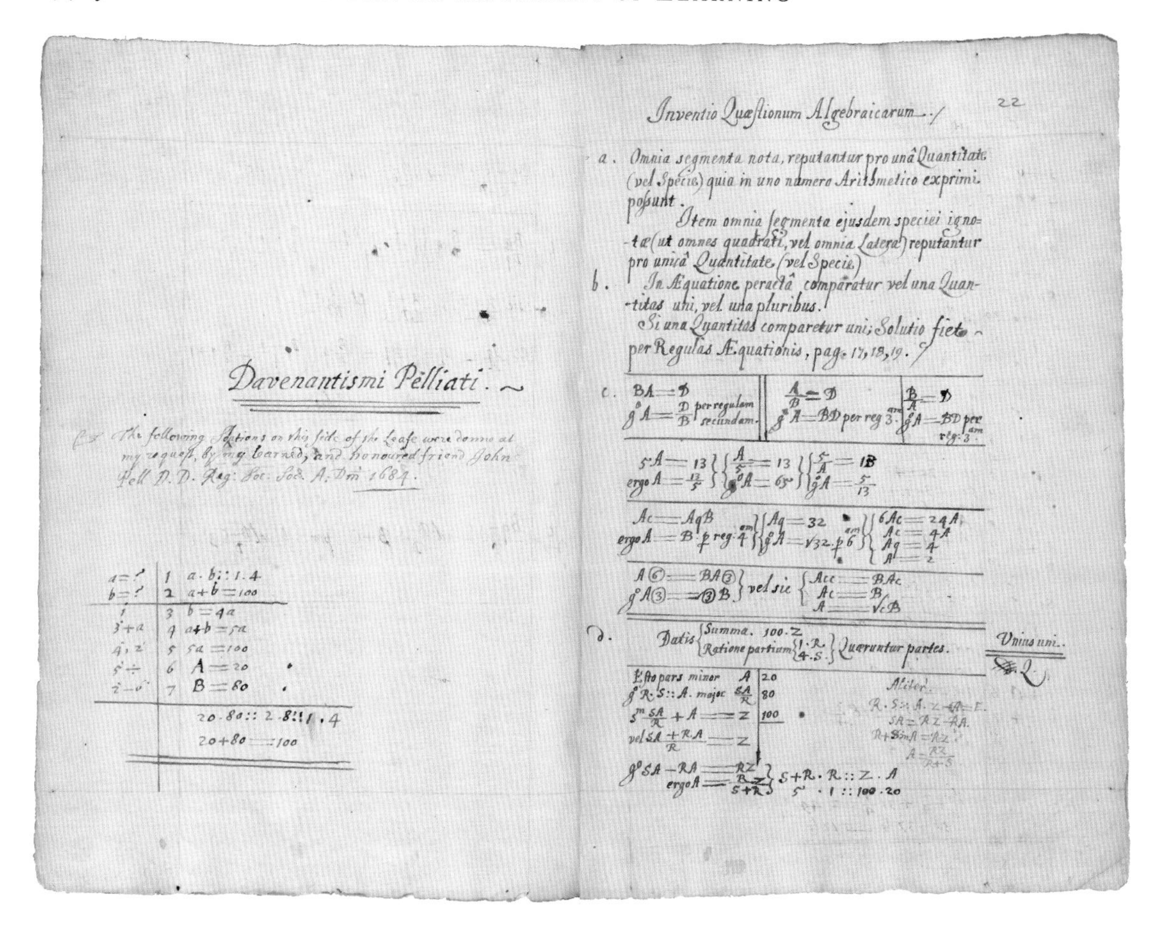


**Figure 27** One of Aubrey's own mathematical notebooks, open at 'Davenantismi Pelliati', Aubrey's copy of an algebraic tract by Edward Davenant with animadversions commissioned by Aubrey from John Pell. Oxford, Worcester College, MS 5.5, fols 21v–22r.

Pellified', a type of expression Aubrey used elsewhere in his mathematical papers) to accompany it; that is, a set of solutions to Davenant's problems calculated by John Pell (1611–1685), an important mathematician whom Aubrey tried to assist in later life, and who may also have taught Aubrey. They certainly discussed mathematics frequently, and Aubrey published Pell's short essay on Roman chronology in his *Miscellanies* (1696: 22–6). Aubrey's chief interest in Pell's mathematics appears to have been the latter's emphasis on method, something which appealed to Aubrey's pedagogical animus. Aubrey also acquired from Pell a manuscript distillation of the problems set out in Leonard and Thomas Digges's Elizabethan textbook *Stratioticos* (1579), a book we have encountered. Pell and Aubrey had much else in common: they were also both pathetic at getting their own works published, and at looking after their own finances. Another important mathematical teacher was Nicolaus Mercator (1620?–1687), an FRS originally from Holstein, who taught mathematics in London, and worked on both theoretical and practical issues, including astronomy and the problems of marine chronology. Aubrey copied out 'his Lessons to me', as well as an acoustic treatise, now MS Aubrey 25. Aubrey's

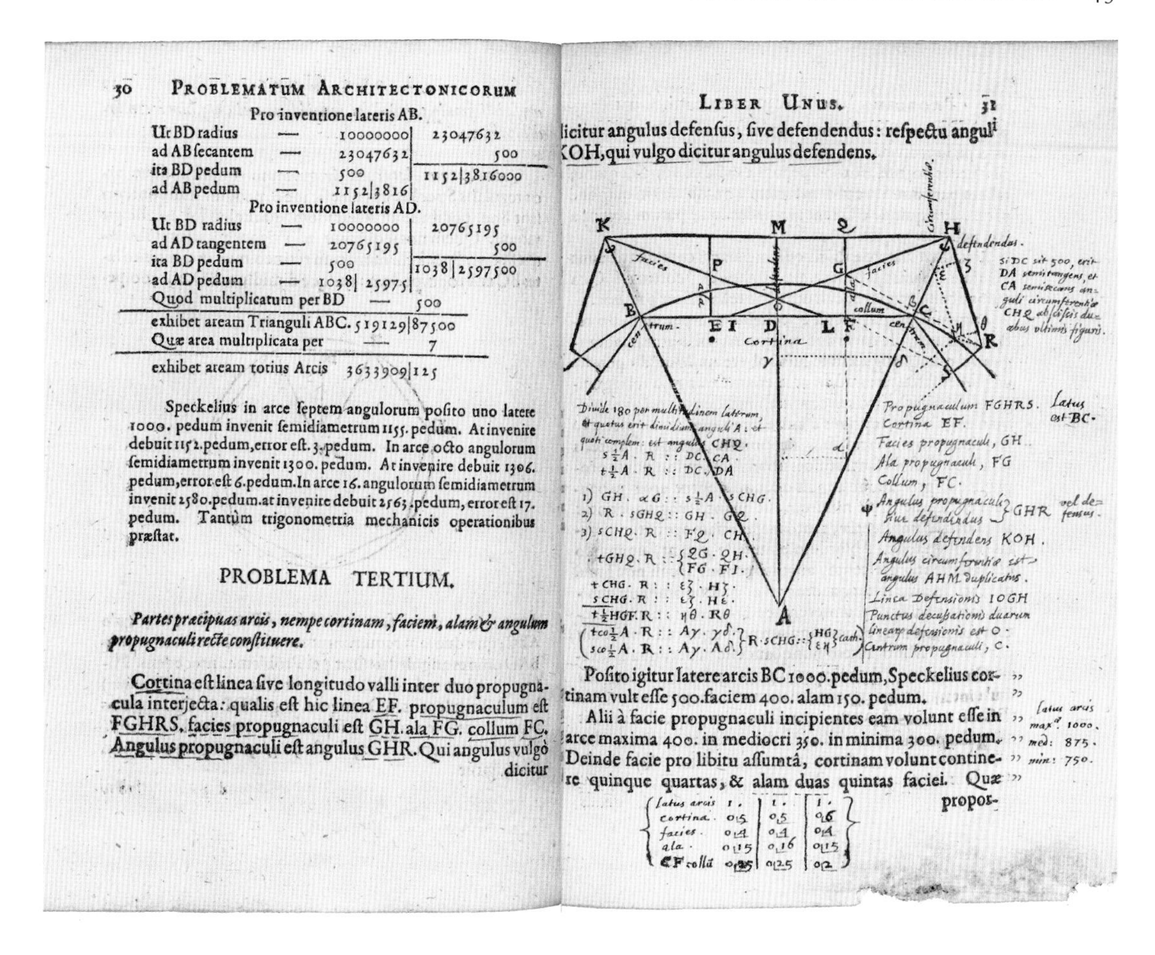

30 PROBLEMATUM ARCHITECTONICORUM

Pro inventione lateris AB.

Ut BD radius — 10000000 | 23047632
ad AB secantem — 23047632 | 500
ita BD pedum — 500 | 1152|3816000
ad AB pedum — 1152|3816

Pro inventione lateris AD.

Ut BD radius — 10000000 | 20765195
ad AD tangentem — 20765195 | 500
ita BD pedum — 500 | 1038|2597500
ad AD pedum — 1038|25975
Quod multiplicatum per BD — 500
exhibet aream Trianguli ABC. 519129|87500
Quæ area multiplicata per — 7
exhibet aream totius Arcis 3633909|125

Speckelius in arce septem angulorum posito uno latere 1000. pedum invenit semidiametrum 1155. pedum. At invenire debuit 1152.pedum, error est. 3. pedum. In arce octo angulorum semidiametrum invenit 1300. pedum. At invenire debuit 1306. pedum, error est 6.pedum. In arce 16. angulorum semidiametrum invenit 2580.pedum. at invenire debuit 2563.pedum, error est 17. pedum. Tantùm trigonometria mechanicis operationibus præstat.

PROBLEMA TERTIUM.

*Partes præcipuas arcis, nempe cortinam, faciem, alam & angulum propugnaculi rectè constituere.*

Cortina est linea sive longitudo valli inter duo propugnacula interjecta: qualis est hic linea EF. propugnaculum est FGHRS. facies propugnaculi est GH. ala FG. collum FC. Angulus propugnaculi est angulus GHR. Qui angulus vulgò dicitur

LIBER UNUS. 31

dicitur angulus defensus, sive defendendus: respectu anguli KOH, qui vulgo dicitur angulus defendens.



Posito igitur latere arcis BC 1000. pedum, Speckelius cortinam vult esse 500. faciem 400. alam 150. pedum.

Alii à facie propugnaculi incipientes eam volunt esse in arce maxima 400. in mediocri 350. in minima 300. pedum. Deinde facie pro libitu assumtâ, cortinam volunt continere quinque quartas, & alam duas quintas faciei. Quæ propor-

Worcester College mathematical transcripts also include materials by John Collins and Sir Jonas Moore, and altogether his collection of short tracts, all by men known to him as teachers or colleagues, covers both theoretical and practical realms.

Aubrey sealed his relations with his old university by two sets of mathematical donations. In addition to his other gifts, he presented to the Ashmolean Museum two instruments on the design of his old hero from Trinity, Francis Potter: first, a quadrant probably made by the local maker John Prujean (which the Keeper, Edward Lhuyd, suspected would be considered rather out of date); and second, Potter's dividing compasses, 'which will divide an Inch into 1000 equall parts'. Alas neither of these instruments can be identified today. But Aubrey's second set of gifts, namely the small mathematical library he presented to Gloucester Hall (now Worcester College) probably in 1695, survives probably in its entirety in Worcester College Library, along with three manuscript collections, mainly containing the short manuscript copies we have just encountered. Why Aubrey chose to donate between forty and fifty items to the Hall is unclear, but it has been suggested that

**Figure 28** 'This was old Mr Oughtreds booke, and the Notes are of his owne handwriting': Aubrey acquires an important copy of Bartholomaeus Pitiscus, *Trigonometriæ siue de dimensione triangulorum* (Frankfurt, 1612). Oxford, Worcester College, pp. 30–31.

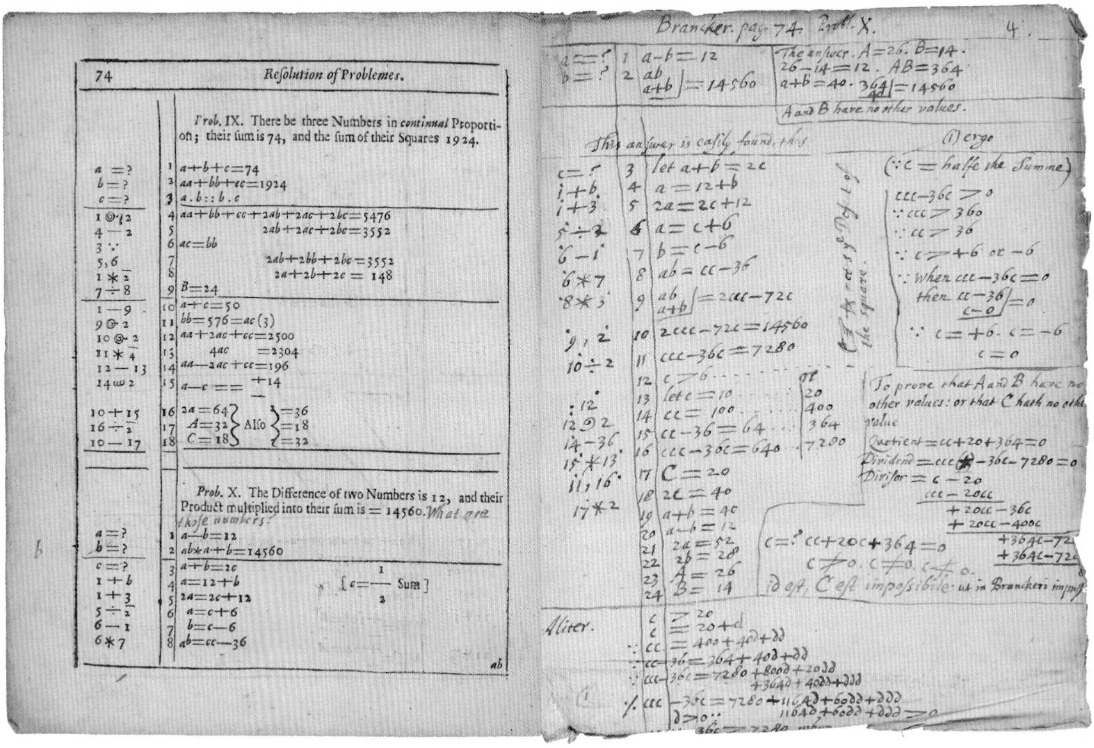

**Figure 29** 'This Book I have carefully corrected by Dr Pell's: and it is to be valued as a Jewell of price': Aubrey's interleaved copy of Thomas Branker, *An Introduction to Algebra* (London 1668). Oxford, Worcester College, p. 74, and facing MS insert.

Aubrey's friend the Principal, Benjamin Woodroffe, was looking for mathematical books to make his proposals for a Greek College at Oxford more attractive. At any rate, two years before his death Aubrey still had such a collection of books to give, perhaps taken from the box marked 'Idea' which he had deposited with Hooke, in which he had placed his 'Idea of Education' manuscript along with a set of appropriate books.

Many of the Gloucester Hall gifts display noteworthy provenances. Works by Mercator, John Gadbury, Thomas Hobbes and Sir William Petty are all presentation copies to Aubrey. The copy of Thomas Digges's *Pantometria* was once owned by Ben Jonson. Bartholomæus Pitiscus, *Trigonometriæ siue de dimensione triangulorum* (1612) is marked by Aubrey 'This was old Mr. Oughtreds booke and the Notes are of his owne handwriting.' His copy of Thomas Branker's *Introduction to Algebra* (1668) has been annotated with solutions by John Pell. Aubrey had long been interested in receiving news of significantly annotated mathematical books. In his life of William Camden, for instance, he noted that his mathematical teacher Mercator owned Camden's copy of Stadius's *Ephemerides*; 'his name is there (I knowe his hand) and there are some notes by which I find he was Astrologically given.' He sought and kept such books, and some were therefore among the set donated to Gloucester Hall.

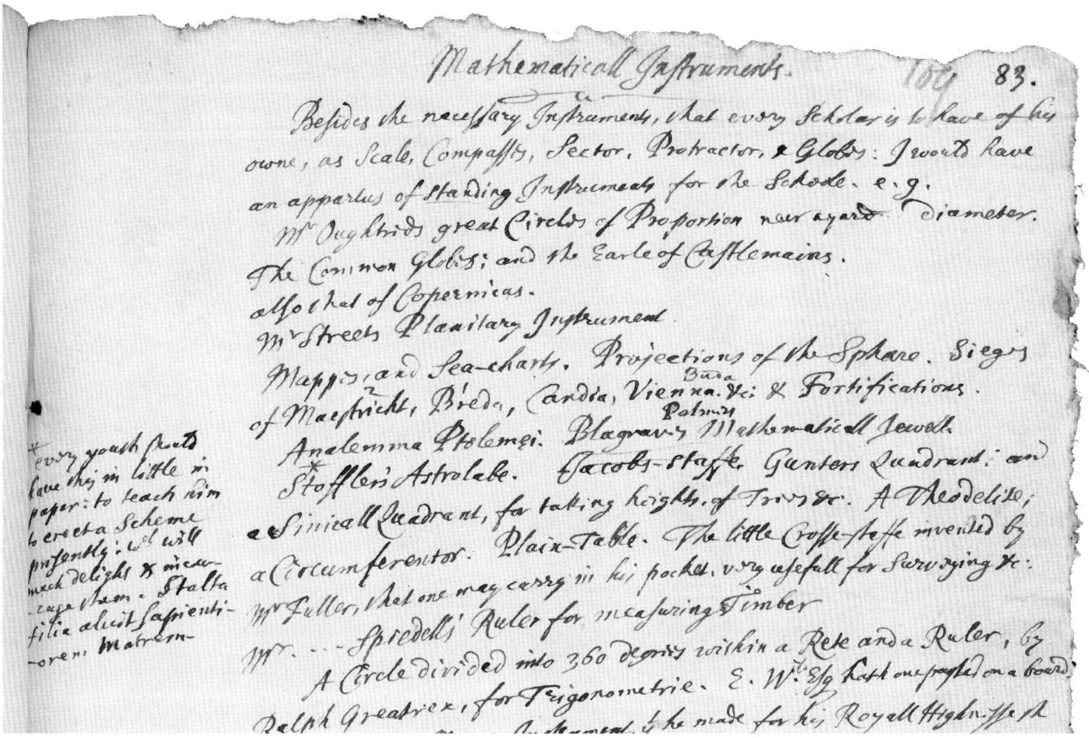

Perhaps the most interesting of Aubrey's mathematical schemes, however, are the sections on the teaching of mathematics that he wrote for his treatise of educational reform, the 'Idea of Education of Young Gentlemen' (now MS Aubrey 10). Aubrey started the 'Idea' in 1669, worked it into a full draft by 1684, but tinkered thereafter for the next decade, investigating the possibility of getting it printed, or of at least generating a master manuscript from which copies could be taken. This manuscript is remarkable from the point of view of mathematical teaching for advocating rather advanced mathematics for schoolboys, and resembles much more the modern curriculum than that of the schools of his time: 'Arithmetique, & Geometrie are the Keys, that open unto us all Mathematicall and Philosophicall Knowledge: and by consequence (and indeed) all other Knowledge, sc. by teaching us to reason aright, and carefully, not to conclude hastily and not to make a false steppe' (Aubrey 10, fol. 8r). The schoolboy should carry around in his pockets some 'Dialogues' for literary style, but also Euclid's *Elements* 'as religiously as a Monke his Breviarie' (MS Aubrey 10, fol. 94r). This kind of instruction, for Aubrey, should displace unnecessary quibbling in logic, or the penning of pointless literary exercises. Although Aubrey's schoolboys are worked hard, they are set to learn their maths following the analytic methods of Descartes, and of Aubrey's friends Pell and Hobbes. In a movement that reminds us not only of Aubrey's youthful

**Figure 30** Aubrey's list of mathematical equipment to be owned by the ideal school. Oxford, Bodleian Library, MS Aubrey 10, extract from fol. 109r.

*Top right* **Figure 31** Quadrant, by John Rowley, *c.* 1700. Oxford, Museum of the History of Science, MHS 31785.

*Middle right* **Figure 32** Circular slide-rule and horizontal instrument, by Elias Allen, *c.* 1634. Oxford, Museum of the History of Science, MHS 40847.

*Bottom right* **Figure 33** An English Camera Obscura, *c.* 1710. Oxford, Museum of the History of Science, MHS 75945.

correspondence with Lydall, but also the educational theories of the Hartlibian reformers, Aubrey stressed that mathematical competence will then lead to 'Mathematicall Prudence', as he titles his chapter on estate management. Mathematics 'makes men most adroict & fitt for Businesse: it makes them to reason aright, & to become Prudent: to be good Lawyers, & Soldiers' (MS Aubrey 10, fol. $41^{b}$r). Aubrey was always interested in mathematical prudence as a means of making money, a further Baconian emphasis. Mathematics, Aubrey says, will make boys 'Fabri suae Fortunae', makers of their own fortunes (MS Aubrey 10, fol. 143r). The outstanding example, as Aubrey immediately notes, is his 'ever honored Friend' Sir William Petty, a man who was as talented at making money as Aubrey was at losing it. (Petty was worth £8,000 a year, Aubrey ruefully notes in his margin.) Aubrey compiled his own personal 'Faber Fortunae' manuscript (MS Aubrey 26), but unlike omnicompetent Petty he proved unable to follow his own advice.

Aubrey's ideal classroom was therefore fitted out with a permanent collection of '*standing* Instruments' for the boys to use; they were also to purchase their own measure, compasses, sector, protractor and globes. The standing collection included some standard devices such as quadrants and astrolabes, but a quantity of instruments invented by Aubrey's own friends or acquaintances: Oughtred's 'Great Circles of Proportion neer a yard diameter', Thomas Street's 'Planitary Instrument' (an equatorium), Hooke's 'portable Instrum*en*t for Surveying', Sir Jonas Moore's universal dial, Jeremiah Grincken's 'Arithmeticall Instrument' (a mechanical calculator actually designed by William Pratt), as well as a plane-table (such as Aubrey had used to survey Avebury), microscopes, telescopes, a theodolite, a 'circumferetor' for measuring horizontal angles, and perhaps most strikingly 'Darkroomes for taking ones picture or Landscap: It will sett them *agog* (as they say)' (MS Aubrey 10, fol. 109r). On the walls, too, there were to be maps, sea charts, and depictions of sieges and fortifications.

Aubrey's ideas in this respect were very much ahead of their time, although, as he explains in his preface, he knew of several similar private academies that were already up and running. Nevertheless, it would be impressive even today to see his rather extraordinary depiction of the ideal maths classroom put into practice, although it must be remembered that Aubrey envisaged his ideal pupils as solely the sons of gentlemen. No doubt Aubrey was also writing down some instruments as they occurred to him, as his list is rather disorganised. But what is perhaps most remarkable is his pedagogical emphasis on mathematical 'delight', a word Aubrey uses frequently – 'that wonderfull, delightfull & usefull Science of the Mathematicks'; '[boys] must be enticed on by pleasure, & delight' (MS Aubrey 10, fol. $29^{a}$r, v). As he wrote earlier in the treatise, 'having espoused a Love to these things, [the pupils] will now willingly undergoe the trouble of extracting Square, and Cube-rootes &c: w$^{ch}$ were they to learne in the first place they would never endure the harshness of it' (MS Aubrey 10, fol. 29r). For these reasons, too, Aubrey thought that rote learning was only admissible if its fruits were carefully explained in advance to pupils. Likewise Aubrey hoped that 'Darkroomes' – he probably means a camera obscura or an arrangement with a scioptric ball – would help out in the classroom not just as instruments of instruction, but as instruments of wonder. Aubrey was a product of a university environment that had fostered a burgeoning interest in the mathematical arts, both theoretical and practical. But not content with this, Aubrey, although no profound mathematician himself, wanted to see that same interest projected back into education at an earlier stage. Aubrey had little but scorn for the grammar-school education he had endured, although he sensibly stopped short of advocating the abolition of Latin in his ideal academy. Nevertheless, his 'Idea' was for Aubrey a sincere plan which he wanted to see put in practice, and he was exasperated by the difficulties he had encountered when trying to secure an aristocratic patron among his friends for such a school: 'if the Nobless have a mind to have their Children be putt in the Clergies pockets, much good may [it] do 'em', anti-clerical Aubrey grumbled (MS Aubrey 10, fol. 2r). But if his ideal curriculum was to be implemented, Aubrey recognised too that the teaching of mathematics had to be enhanced by finding a better method, so that facility is more swiftly attained; and by exploiting the delight of mathematics, so that learning it was fun.

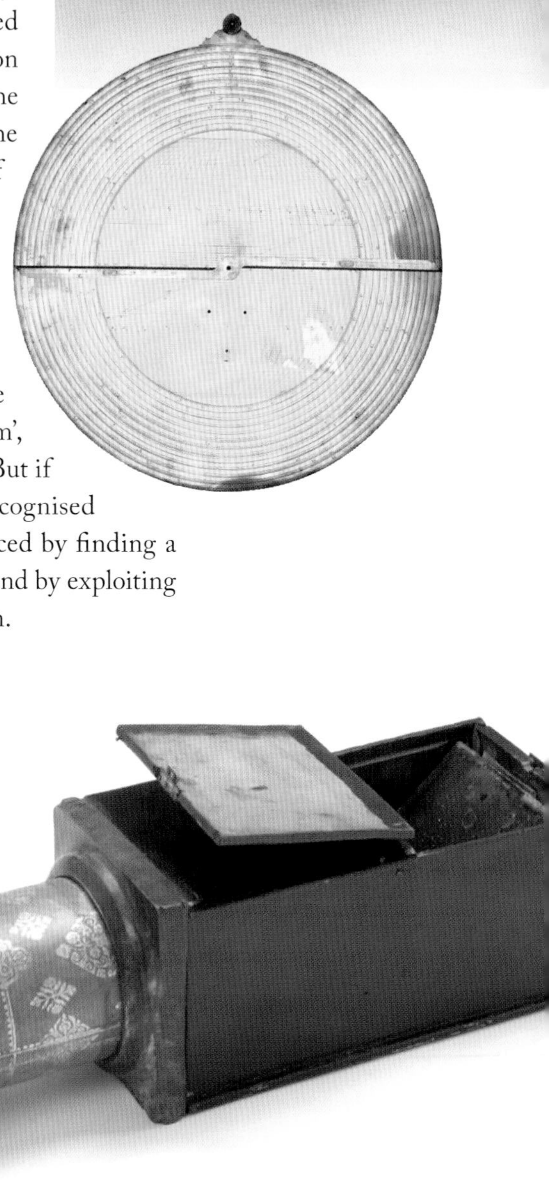

# 5

# The Philosophical Language

Aubrey was well aware that he possessed two powerful personal assets: he was amiable, and he was extremely well connected. This made him reflect in his autobiographical notes that his worth lay in being an intellectual fixer, a man who could broker and then orchestrate intellectual encounters between scholars otherwise unknown to one another. Quoting Horace, Aubrey described himself as a '*Cos ... exors ipse secandi*' – a 'whetstone' for sharpening other objects, 'though itself unable to cut'. Aubrey immediately followed this typically modest appraisal by remembering the intellectual collaboration organised by him of which he was most proud: 'e.g. universall character which was neglected and quite forgott and had sunk had not I engaged in the work' (*BL* 1.39).

The 'universall character' of which Aubrey speaks was the attempted creation in the later seventeenth century of a fully workable artificial language, a new and better method of communication based not on extant natural languages, but created entirely afresh, designed by philosophers, and possessing a rational structure, a new script and a new pronunciation. The fabrication of an artificial language which was not simply a distillation of extant languages but itself an embodiment of philosophical truth and the new science is one of the most fascinating intellectual projects of the seventeenth century, and Aubrey was right to recognise that he had played a major role in orchestrating its final phase. The project failed, as all such projects must, but its life cycle gets us closer to the heart of seventeenth-century intellectual aspiration than any other movement, and Aubrey was part of it throughout its three climactic decades.

Why did Aubrey's age feel the need to create a new, 'philosophical' language? Various motives were claimed, and the value of a shared international language for the business of religious conversion and unification ('Babel revers'd') was a benefit unsurprisingly touted by the 'language planners', as they have been usefully dubbed. As John Wilkins, by now Bishop of Chester, claimed in the preface to his own celebrated *Essay towards a Real Character, and a Philosophical Language* (1668), the largest and fullest attempt in the period, his philosophical language would remedy the Curse at Babel, dispense with the need to learn foreign languages, facilitate commerce, spread true religion, and indeed reform religion itself – 'by unmasking many wild errors that shelter themselves under the disguise of affected phrases; which being Philosophically unfolded, and rendered according to the genuine and natural importance of Words, will appear to be inconsistencies and contradictions' (1668: sg. b1r). No doubt the Bishop of Chester meant what he said. But the dominant impetus behind the project was rather a shared sense that the rise of experimental philosophy as a self-consciously new activity deserved a new kind of language; and conversely an attendant dissatisfaction with the traditional Aristotelian philosophy was bound to result in an attack on its vocabulary, an ossified Latinity. And if a better language furthered better theology, better education and better trade, then who could object to such an innovation?

Interest in the possibility of what was termed a 'real character' had been sparked off in English philosophy by Francis Bacon, who had remarked that the advancement of learning was in practice an attempt to overcome the two curses delivered by God in Genesis: the curse of labour following the fall of Adam and Eve, and the curse of the division of tongues at Babel. The young John Wilkins picked up on this Baconian passage in his still highly readable manual on codes, *Mercury, or the Secret and Swift Messenger* (1641). The first curse, he proposed, was continually being counteracted by advances in 'Arts and Professions', or technological progress. But the second curse, he contrasted, had been only partially alleviated by the institution of Latin as the international language of scholarship. Would it not be better, challenged Wilkins, if an entirely new script could be developed? – in which 'primitive words' were each assigned a special character, with conjugations and declensions expressed by special marks set around this 'radical' symbol. At this point Wilkins was envisaging a universal *script*, not a spoken language; the effable component would come later. He also explicitly assumed that 'though severall Nations may differ in the *expression* of things, yet they all agree in the same *conceit* of them' (1641: 105–10). In other words, Wilkins assumed that all peoples in all places at all times think in fundamentally the same way, and that this homogeneity of thought could be tapped directly using an appropriately constructed language, and then distributed universally through an appropriately designed script. As we shall see, it was the challenge to this assumption posed by the philosophy of John Locke in the 1690s that sealed the doom of the already waning artificial language movement.

Wilkins issued his challenge in 1641, but the idea of an artificial language was already being discussed, especially in the circle of the London intelligencer Samuel Hartlib. Many of the earliest comments on such a language were inspired by contemporary hypotheses about the nature of Hebrew, Egyptian hieroglyphic, and the recently discovered Chinese and Mesoamerican scripts. Hebrew, a language by this point widely studied in European universities, generated a large vocabulary

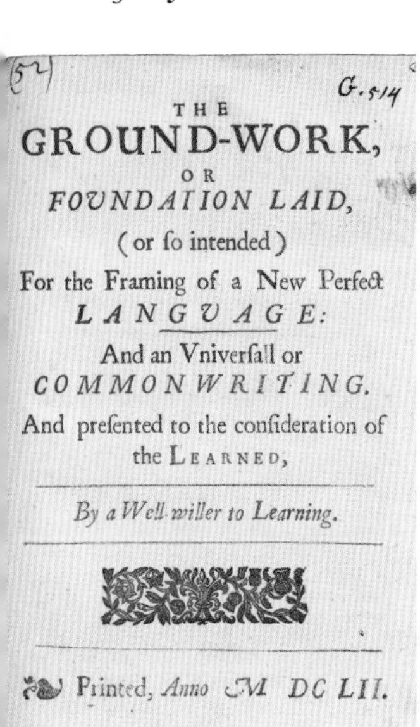
THE
GROUND-WORK,
OR
*FOVNDATION LAID,*
(or ſo intended)
For the Framing of a New Perfect
*LANGVAGE:*
And an Vniverſall or
*COMMON WRITING.*
And preſented to the conſideration of
the Learned,

*By a Well-willer to Learning.*

Printed, *Anno* M DC LII.

**Figure 34** Francis Lodwick, *The Ground-Work* (London 1647). The alphanumeric code in the top right-hand corner shows that this book once belonged to Sir Hans Sloane. Oxford, Bodleian Library, 4° E 3(20) Jur, title page.

from a comparatively small number of radicals; whereas the other three scripts, not understood at all by the early moderns, appeared to have bypassed the notation of sound altogether, going straight for ideographical representation of the thing signified. Although the early moderns were mistaken about the real nature of Chinese and hieroglyphic script, they were at least rightly suspicious of totally ideographical languages, realising that a one-thing, one-symbol language would result in an impossibly large lexicon. An artificial language, therefore, could not simply be a list of things. The language planners would have to find a process whereby reality could be reduced to its fundamental 'radicals', either of substances, or of thoughts, or of both. Then complex things or ideas could be expressed as a combination of simpler elements, or as located in a certain organised and organising hierarchy. Once a script had been developed to express such a hierarchy, then a form of speech isomorphic to that script could be fitted to it.

Aubrey entered into this discussion very early. In December of 1652, Samuel Hartlib wrote down in his diary that Aubrey had reported that Potter was 'studying a common or univ[ersal] character', and a year later Aubrey told Hartlib that Potter was 'about a com*m*on-Language to bee written in ordinary Characters' (*HP* 28/2/43A; 28/2/80B). Aubrey will have picked up his interest primarily from Potter, then, but there were other sources too. The great inventor and statistician William Petty, later to become, as we have seen, one of Aubrey's most admired and envied friends, had dreamed in 1647 of an ideal new school in which the pupils 'be not onely taught to Write according to our Common Way, but also to Write Swiftly and in Reall Characters' (1647: 5). In that year too appeared the first formal proposal for such a language, *A Common Writing whereby two, although not understanding one the others Language, yet by the helpe thereof, may communicate their minds one to another*, devised by a London merchant of Flemish–French extraction, Francis Lodwick. Lodwick was probably the only man in the seventeenth century whose interests in artificial language spanned a lengthier period of time than Aubrey's own, and once again Aubrey and Lodwick would later become friends, working together on the revision of Wilkins's *Essay*. Lodwick's short work was a proposal merely for an international script, divided into three sections: the first proposed a universal grammar; the second set out a script adapted to that grammar; and the third explained a syntax for the script, which was to be mounted on a stave rather in the fashion of musical notation. Then in 1652, the year of Aubrey's first documented interest in the project, Lodwick produced a second, even shorter, but more ambitious proposal, *The Ground-Work or Foundation Laid (or so intended) for the Framing of a New Perfect Language*. This was to be a language that was not only written, but spoken, and that would supplant natural languages. It was to consist of a rational grammar and a new notation, itself designed to express its relation to a governing dictionary of all things, the mammoth work of a 'sound Philosopher'. This lexicon would be accompanied by a second book, a 'Law', namely the full grammatical system sketched out in brief in *The Ground-Work* itself.

Lodwick's two tiny pamphlets represent the only significant printed proposals to be generated in the earlier phase of the project, but the hum of speculation continued throughout the decade, culminating in Oxford in the three years before the Restoration. This was triggered by the arrival of the Aberdonian George Dalgarno, who had initially been encouraged by Hartlib. Dalgarno arrived at his own ideas by combining two seemingly distinct areas of speculation – Hebrew grammar and shorthand. The devout Dalgarno saw the hand of God in the structure of Hebrew, and realising that shorthand could be adapted to a radical-and-diacritic model, he saw a way towards transforming a technique for swift writing into a technique for philosophical description. Dalgarno was soon taken under the wing of the all-powerful John Wilkins, at that

**Figure 35** The Creed, as translated into John Wilkins's philosophical language, from his *Essay towards a Real Character, and a Philosophical Language* (London 1668). Oxford, Bodleian Library, L 2.14 Art, part of the opening pp. 404–5.

404 *Concerning a Real Character.* Part IV.

*The Creed.*

1 2 3 4 5 6 7 8 9 10 11
I believe in God the Father Almighty Maker of Heaven and
12 13 14 15 16 17 18 19 20 21 22
of Earth, and in Jesus Christ his Son only our Lord, who was
23 24 25 26 27 28 29 30 31 32
conceived by the Holy Ghost, born of the Virgin Mary, suffered
33 34 35 36 37 38 39 40 41 42
under Pontius Pilate, was crucified, dead, and buried, he descended
43 44 45 46 47 48 49 50 51 52 53 54 55 56 57 58
into hell, he did rise from the dead in the day third, He ascended
59 60 61 62 63 64 65 66 67 68 69 70
into heaven, where he sitteth at the right hand of God the Father,
71 72 73 74 75 76 77 78 79 80 81 82 83 84 85 86
from whence he shall come to judge the Quick and the dead. I believe
87 88 89 90 91 92 93 94
in the Holy Ghost, the Church holy Catholick, the Communion of
95 96 97 98 99 100 101 102 103 104 105 106 107

Chap. II. *Concerning a Real Character.*

1. (ʹ|) This Particle being a single point, must denot[e] demonstrative Pronouns in the singular Number: And [...] towards the upper end of the Character, and before the Nominative Case of the first Person, *viz. I.*

2. (o|) This Particle being a Round, must signifie the [...] being placed towards the middle, and joyned with the [...] case of the first person, it must be rendered (*Am.*)

3. ( ) This Generical Character is appointed to [...] Genus of *Habit*. The affix on the left hand making a right [...] the lower side, doth signifie the fifth Difference under [...] which according to the Tables, is *Infused Habit*. The othe[r] [...] king an obtuse angle, doth signifie the fourth Species, [...] which is defined to consist in a readiness of mind to yield [...] assent (without any affected captiousness) unto revealed [...] on such Grounds as their natures are capable of, and su[...] themselves sufficient to prevail with a prudent teachable [...] Hook upon the Difference denotes the *Active* voice, an[d] [...] upon the Species, the *Adjective*, viz. *Believing*.

The Word *Believe* being of an equivocal sense, may like[...] pressed by the Genus of *Spiritual Action*, the second differen[ce] [...] to the Acts of the *Speculative Judgment*, and the affinis of [...] Species, which is *Believing*, described to be an assent to a[...] truth upon the credit of others by whom it is related. B[...] chuse the former, as being more proper to this place.

4. ( ) A streight Line being the most simple, is put fo[r] [...] racter of *God*. The acute angle on the left side doth deno[te] [...] Person of the Blessed Trinity, namely, *God the Father*; wh[...] lowing the Verb, is supposed to be in the *Object case*. An[...] better express the true notion of *Credo in Deum*, than by usi[ng] [...] position *In*, as the sense of that Particle is determined in the Tables.

5. ( ) This Character doth by the two strokes d[...] *Hyphen*, appear to be a *Compound*. The two Points denote [...] of the third Combination, and by the place of them at th[e] [...] they must signifie the *Universal Collective*, viz. *All*. Th[e] [...] Character with which this is compounded, is the Genus [...] *Power*. The Hook at the end signifies the Grammatical not[...] *jective*. So that this compound Character may be rendered A[...] ful, Omnipotent, or *Almighty*.

6. (ʹ|) This Particle doth signifie the *Demonstrativ[e]* *The*.

point Warden of Wadham College. The two men collaborated, but, in a prolepsis of what was to happen within Aubrey's own orchestra in the 1670s, they soon realised that they differed profoundly about the basis of the envisaged language, and they fell out. Perhaps nervous that Wilkins would eclipse his own work, Dalgarno therefore rushed out in 1661 his *Ars signorum, vulgo character universalis et lingua philosophica* or 'The Art of Signs, commonly called the universal character and the philosophical language'. This was prefaced by an address to Charles II written in Dalgarno's language, followed by a Letter of Recommendation signed by Charles himself, and attested by men including Wilkins, Seth Ward and John Wallis. Evidently, therefore, Wilkins was still willing to put in a good word for Dalgarno, even if he would not work with him. Nevertheless, Wilkins pointedly refused to mention Dalgarno by name in his own *Essay*, and later in life Dalgarno vented his frustration in a private manuscript autobiography, clearly still smarting at how easily the great academician–churchman–politician Wilkins could cast off an obscure Scots schoolmaster. Dalgarno's work, however, was to be carefully read by the philosopher Leibniz, and his claim that his system was more logically advanced than Wilkins' essentially classificatory venture is not without some justice.

The work which everyone read, however, was John Wilkins's imposing folio *Essay* (1668). Wilkins had been working on his magnum opus for a decade and a half by this point, in collaboration with many other leading scholars, and a wide reception could be guaranteed. Wilkins opened with a history of natural languages, all derived from Adamic Hebrew, and an analysis of their defects. He then presented a huge 'Universal Philosophy', essentially a classification of all things, then a 'Philosophical Grammar', and finally the real character itself, to be matched to the grammar, with instruction on its notation, and how it was to be pronounced. The book, despite the disastrous loss of much of the initial printing in the Great Fire, was a typographical triumph, featuring not only engraved plates of Chinese and a fold-out illustration of Noah's Ark, but the real character itself, set from type specially designed by the printer Joseph Moxon, who a decade later would become the first tradesman to be elected to the Royal Society. Wilkins's *Essay* proper was then followed by a dictionary compiled by William Lloyd, the future Bishop of Worcester and keen speculator on the Millennium. This huge work, published under the Society's own imprimatur, was eagerly read, and despite Wilkins' careful protestations that this was only an essay *towards* an adequate language, and that he both required and desired further correction, for readers here at last was a system that could actually be put into practice. If I wish, for instance, to express the notion 'sneeze', I consult first the tables in part two, where I find that 'sneezing' is classified under the fourth type of 'Motion', namely 'Purgation', of which sneezing is the first type. The fourth part tells me how to notate this idea: motion is a subclass of action, and is notated by the 'genus' marker of a horizontal bar with right-facing hook mounted upon it: ⌐. As purgation is the fourth 'difference' of motion, I add the sign ⟋—. And as sneezing is the first 'species', I finish off the character with the first 'species' marker —⇀, constructed as mirror-images of the 'difference' markers. (Ironically, this genus–difference–species terminology betrays just how deeply embroiled in Aristotelian logic and its world Wilkins still was.) My complete character for a sneeze, therefore, consists of a genus–difference–species combination character. And if I know my script well enough, I can decode from any given character exactly where it comes from in the hierarchy of the language, and hence the script tells me about the philosophical place of the thing, idea, action or relation notated. It is in this sense that the language is 'philosophical' – just from the shape of the genus marker I can discern at once whether I am dealing with God, a stone, a shrub, a judicial concept, a naval concept or some type of habit or motion. If I want to start forming complete statements, I then need to turn to part three, in which the grammar of the language is explained. And in practice the real character is a language in which one can write relatively complex things. I can even notate irony, by following a statement with an upside-down question mark.

Certainly Wilkins and his readers experimented with trying to communicate in the proposed script. Just after his work had been published, Wilkins wrote to the FRS and Savilian Professor of Geometry John Wallis in his script, and Wallis replied, placing a transcript of their correspondence in the front of his own copy of the *Essay*. Wilkins eventually deposited this copy in the Savilian Library, and hence it is in the Bodleian today, still with the Wallis and Wilkins inserts. Nor was this the first time Wallis had gone to the trouble of learning how to work with an artificial language, for the Bodleian copy of Dalgarno's *Ars Signorum* once owned by Wallis bears his handwritten interlinear translation of Dalgarno's epistle to the King in his own language: 'Shod Caroli kanel sʋfa, tηn sefap, sηf Brittanoi sʋma, Hibernoi, Frankoi; tηn Krumel suf Tʋpu Siba Christoisa' – that is, 'To Charles the king most great & most potent of Brittain great, Ireland, France and defender of faith true Christian', as Wallis translates, word for word. (And just in this little snatch we can spot the superlative and then the comparative form of the same word here in accusative form: 'sʋfa'/'sʋma' ('sʋm', 'sʋn', 'sʋf' = 'rather', 'more', 'greatest'); Dalgarno in particular criticised Wilkins's overcomplex treatment of comparatives.) The Ipswich schoolmaster Cave Beck, who himself published a very crude artificial

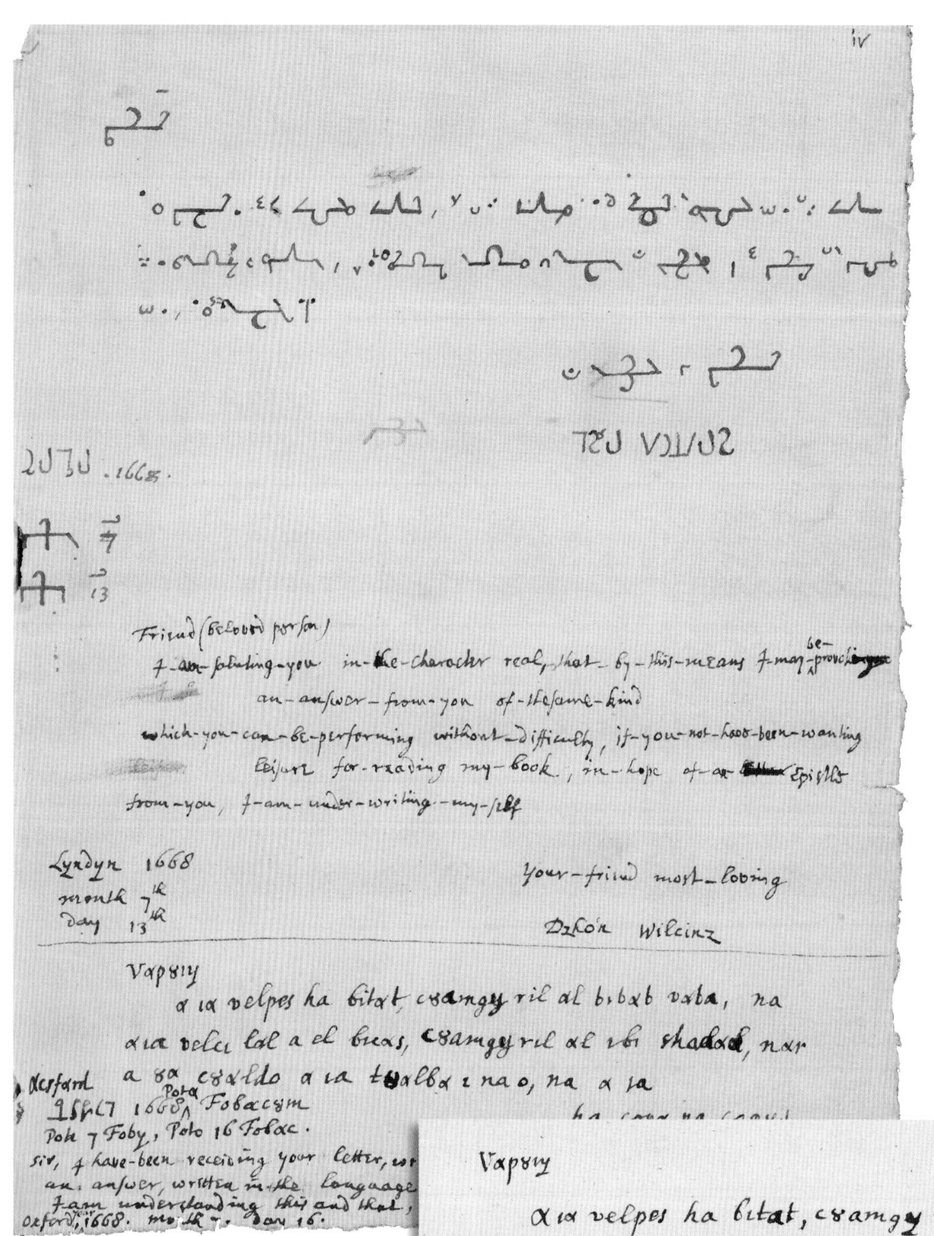

Friend (beloved person)

I-am-saluting-you in-the-character real, that-by-this-means I-may-be-provoking an-answer-from-you of-the-same-kind which-you-can-be-performing without-difficulty, if-you-not-have-been-wanting leisure for-reading my-book, in-hope of-an Epistle from-you, I-am-under-writing-my-self

Lyndyn 1668 month 7th day 13th

Your-friend most-loving

Dzhon Wilcinz

**Figure 36** John Wallis and John Wilkins correspond in the philosophical language, preserved in John Wallis's copy of the *Essay*. Oxford, Bodleian Library, Savile A4, fols iv, vi.

language in 1657, sent a letter to Robert Hooke in 1676 in Wilkins's character; and the next year Paschall in Somerset wrote to Aubrey in parallel English/Wilkins, although in truth Aubrey found Wilkins's script difficult to read without help. For this he turned to Francis Lodwick, said to be particularly expert in the real character, reading it off, so Aubrey admiringly commented, *stans pede in uno*, 'standing on one foot' (MS Aubrey 13, fol. 135r). We know that Lodwick wrote a set of private theological essays 'in one of my new Carraters' (MS Sloane 859, fol. 74r); and he also demonstrated the feasibility of Wilkins's language by translating Sir William Petty's *Discourse concerning the use of Duplicate Proportion* (1674) into the real character. Neither of these feats survives today, and Petty did comment drily on the latter venture, 'I take Mr Lodwicks paynes to put that discourse into ye reall character to be an honor to Bp Wilkins & my selfe, but doubt of its acceptance in ye world' (MS Aubrey 13, fol. 100r). Paschall also wrote to Lodwick on alchemical matters in his own phonetic script, itself an adaptation of Wilkins's phonetic script, thereby deploying what was in origin a script developed for scholarly clarity as a handy means of encrypting letters between friends (Royal Society MS Early Letters P, nos 58, 60). Aubrey too recognised the cryptographic potential of Wilkins's real character, making a note to himself in one of his manuscripts on an agricultural 'secretum' he had found out that he 'desire[d]' Lodwick 'to translate this secret into the Universal Language' (MS Aubrey 2, fol. 97v), presumably in order to stop others from profiting from it. Most strikingly of all, Robert Hooke published in his *Description of Helioscopes* in 1676 a text on watch construction in Wilkins's character, a complex gesture simultaneously advertising Wilkins's script, displaying Hooke's proficiency in it, inviting the reader to attempt a translation, but also asserting Hooke's own intellectual ownership of his mechanical innovations. These examples of the decay of systems of writing originally developed for the purpose of widening communication into mere codes is one sign of the almost insuperable social problems facing those wishing to advocate totally new ways of writing.

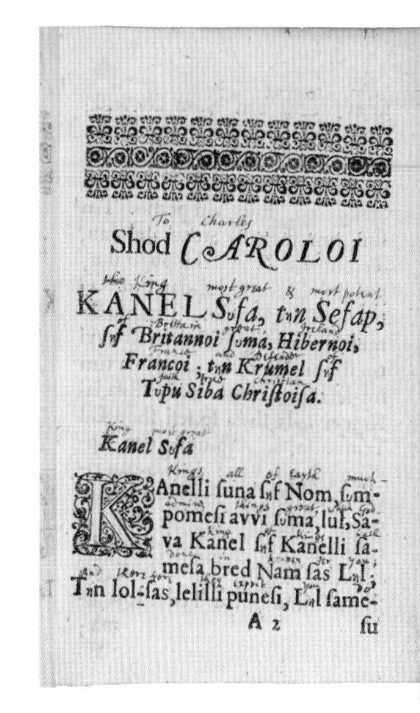
Shod CAROLOI
KANEL Sυfa, tηn Sefap, sηf Britannoi ſuma, Hibernoi, Francoi; tηn Krumel sηf Tυpu Siba Chriſtoiſa.

Kanel Sυfa

KAnelli ſuna sηf Nom, ſumpomeſi avvi ſuma lul, Sava Kanel sηf Kanelli ſameſa bred Nam ſas Lηl. Tηn lol-ſas, lelilli puneſi, Lηl ſameſu
A 2

**Figure 37** John Wallis translates George Dalgarno's preface to Charles II, from his own copy of Dalgarno's *Ars Signorum* (London 1661). Oxford, Bodleian Library, Savile CC 18, sg. A2r.

Indeed, the parallel developments of several scripts, some using entirely new forms, others adapting existing Roman and Greek characters, of 'real' characters alongside simply phonetic scripts, of private versus public uses for such scripts, sometimes for clarity, at other times for obscurity, show that the language planners as a group were at best undecided and at worst confused about the nature, scale and ambition of their task. The Royal Society had itself formally appointed a committee to oversee the revision of Wilkins in 1668, but it failed to make a report. Not that various members of the Society didn't care: even before the publication of Wilkins, Hooke, as we have seen, was

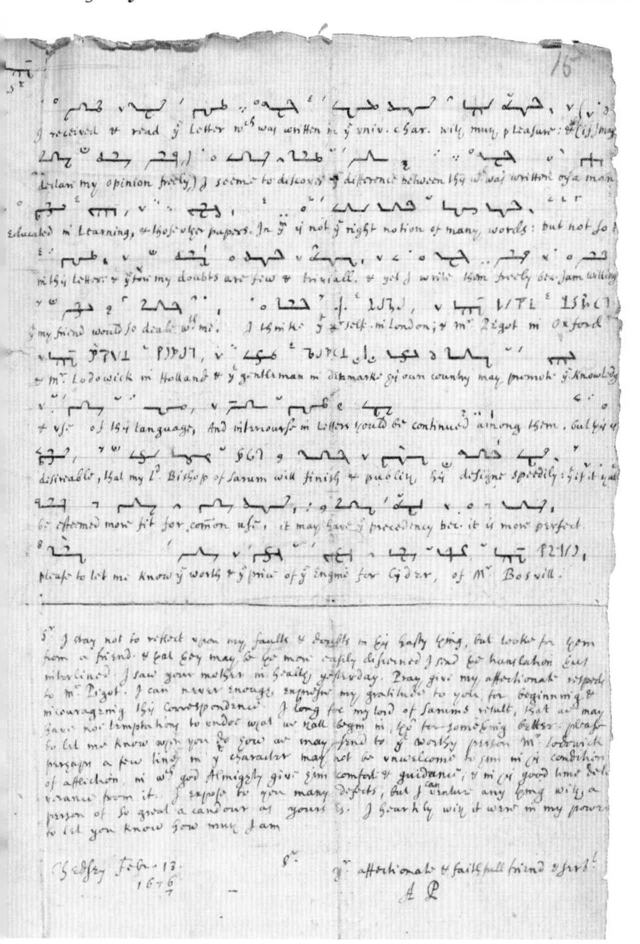

I received & read ye Letter wch was written in ye univ. char. with much pleasure: & (if I may
declare my opinion freely) I seeme to discover ye difference between this wch was written by a man
educated in Learning, & those other papers. In ye is not ye right notion of many words: but not so
in this Letter: & therefore my doubts are few & triviall. & yet I write them freely bec. I am willing
ye my friend would so deale wth me. I think yt yr selfe in London; & Mr Pigot in Oxford
& Mr Lodowick in Holland & ye gentleman in Denmarke in his own country may promote ye knowledge
& use of this language, and intercourse in Letters would be continued among them. but this is
desireable, that my Ld Bishop of Sarum will finish & publish his designe speedily: & if it shall
be esteemed more fit for common use, it may have ye precedency bec. it is more perfect.
please to let me know ye worth & ye price of ye Engine for Cider, of Mr Bosvill.

Sr I stay not to reflect upon my faults & doubts in this hasty thing, but looke for them from a friend. & that they may be the more easily discerned I send the translation thus interlined. I saw your mother in health yesterday. Pray give my affectionate respects to Mr Pigot. I can never enough expresse my gratitude to you for beginning & incouraging this Correspondence. I long for my lord of Sarums result, that we may have noe temptations to undoe what we shall begin in, tho for something better. please to let me know when you doe how we may send to yt worthy person Mr Lodowick perhaps a few lines in ye character may not be unwelcome to him in his condition of affliction, in wch god Almighty give him comfort & guidance, & in his good time deliverance from it. I expose to you many defects, but I can venture any thing with a person of so great a candour as yours is. I heartily wish it were in my power to let you know how much I am

Chedzey Feb. 13. 1676/7

Sr

yr affectionate & faithfull friend & servt

A P

Figure 38 Andrew Paschall writes a letter to Aubrey in Wilkins's language, and offers his own interlinear translation. Oxford, Bodleian Library, MS Aubrey 13, fol. 15r, letter of 13 February 1676.

said to be organising the Society's repository along the lines of Wilkins's language; and in 1674 Aubrey himself said to Wood that he was compiling the repository catalogue in conformity with Wilkins's classifications too. And so it fell to Aubrey in the period around and following Hooke's *Helioscopes* to mobilise an informal band of natural philosophers to revise Wilkins's *Essay* with a view to producing a second edition, this time transacted by a group of philosophers, and not just one man. This, indeed, was just as Wilkins had desired his work to be developed. The original *Essay* itself had been a somewhat collaborative venture. Wilkins acknowledged that his treatment of pronunciation and orthography had drawn on papers lent to him by Lodwick and by William Holder (1616–1698), FRS and brother-in-law of Christopher Wren; and for the tables upon which his language relied, Wilkins turned to Samuel Pepys for naval, and to John Ray for botanical, classifications. Ray was even imposed upon to translate the whole *Essay* into Latin for the international market, a labour he begrudgingly completed, but without the result ever appearing in print; Aubrey's later attempt to get the *Essay* translated into French likewise proved abortive.

Aubrey's revision group likewise divided up the various tasks facing the revisers. The group, geographically scattered and united only as a network of correspondents, consisted of Aubrey himself, the veterans Francis Lodwick and Seth Ward, John Ray and Robert Hooke, and Andrew Paschall and a young Oxford don at Wadham, Thomas Pigot. Paschall was the only one of these men who was not or would not become an FRS; but as a whole this epistolary network shows how configurations of scholarship could be erected on the peripheries of a learned society, sharing in and drawing from that society's intellectual interests, but possessing an identity that was quite distinct from that of a mere subcommittee. Lodwick worked on his old interests, especially phonetics, and he had been circulating a draft phonetic alphabet since as early as 1673. He also undertook the expansion of Wilkins's tables. Wilkins had commented that every natural language is replete with terms to do with rank, office, job, sect, legal process, heraldry, food, drink, clothing, games and music, that an artificial language would

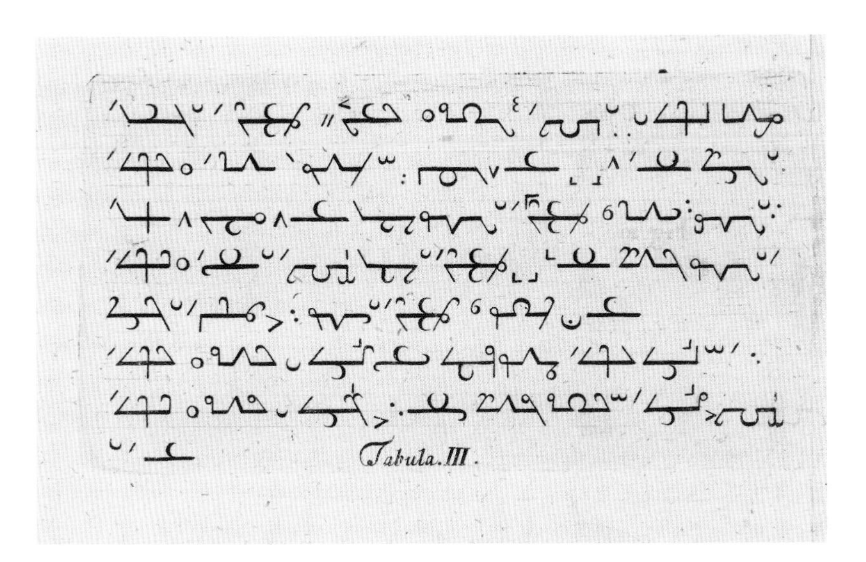


**Figure 39** Hooke encodes his secret of spring watches in the real character and publishes it in his *Helioscopes* (London 1676) in order to 'register' his priority. Oxford, Bodleian Library, Ashmole 1652(3), tabula III.

find very hard to encompass; so Lodwick, himself by profession a merchant, in the only surviving scrap of this work, set about trying to develop such a table for the various English mercantile and political roles. Paschall also worked on phonetic notation, and thought too about educational ploys, such as wall-maps to be hung in greenhouses, so that horticulturalists could see at once how to classify and how to notate their plants. Three of these large botanical posters have survived. Ward had published some extremely interesting remarks on artificial language as far back as 1654, and was invited to develop his theoretical recommendations; while Pigot, the youngest of the group, seems largely to have offered criticism, and to have kept an eye on the general coherence of the group's efforts.

Only a fragment of all this labour is visible today, but it is thanks to Aubrey's careful archiving of his incoming correspondence that we can follow the group's progress at all. Unfortunately very few of the letters of Lodwick and Ward seem to have survived, but by listening to the reactions of especially Paschall and Pigot, we can reconstruct what happened. Perhaps inevitably, it is a story of growing disagreement, even confusion, about what a revised Wilkins was supposed to accomplish. At first, the group hoped that the basic structure of Wilkins's approach could be left as it was, and that it was simply a matter of revising the tables to make them more precise and more concise; and of tinkering with strictly peripheral concerns such as a phonetic script. Paschall, Pigot and Lodwick all agreed that the language as it stood required far too much compounding, and much of the preliminary work, no doubt assisted by Lodwick's practical experience of translating an entire Petty tract into the real character, focused on developing its vocabulary. But it soon became obvious that what was really in question for some revisers was the philosophical substructure of the language. Paschall and probably Aubrey too took some time to grasp what was more obvious to Ray, Lodwick and Pigot, and especially Ward, whose own radically different ideas were only slowly disseminated to the others. Aubrey's own interests were largely in what the language, once perfected, would offer, and how it might fit into the larger question of educational reform. His own prized 'Idea of Education of Young Gentlemen', latterly stored in a chest in Hooke's keeping marked 'IDEA', had promised that the 'Real Character' would assist 'Real Religion', merely a repetition of Wilkins's own blurb, but Aubrey recognised too that there was no future for the language unless it was 'taught to boies'. For Aubrey, then, the real character was not an ancillary language developed for adult scientists, but a system that could and should stand alongside the vernacular in the education of children. Paschall was perhaps the least focused of the group. In 1676, when hopes for the language were still running high, Paschall commented that Wilkins's tables, when finished, would be a great help towards learning Latin, Greek and other languages, scarcely the point of the enterprise. He also praised the language in the same year as a mnemonic tool for students, 'a repository for the matters of their reading & observation'. By 1680, Paschall was still touting the language's use for structuring commonplace books, and in general Paschall's enduring interest was in the language's taxonomic power, and not really in its philosophical structure.

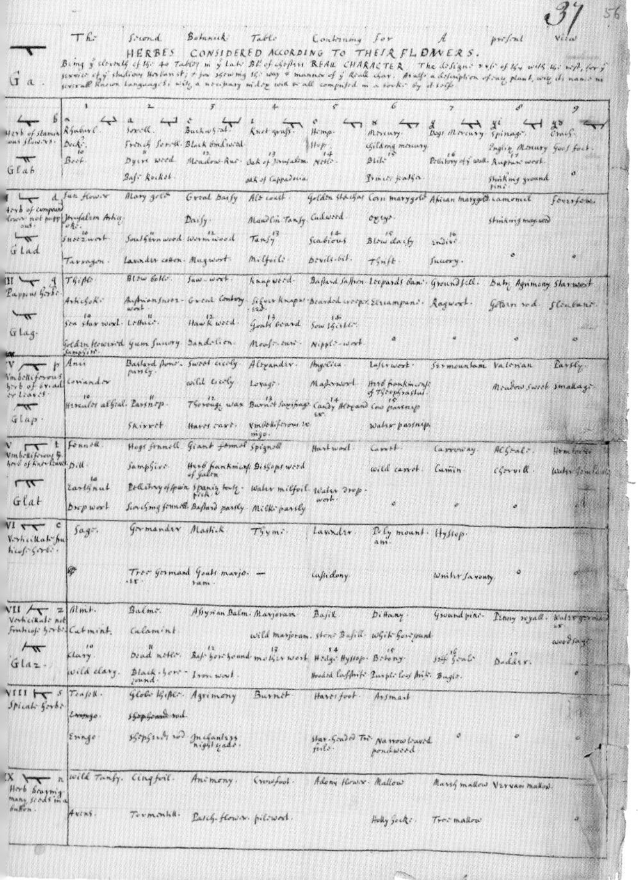


**Figure 40** How to memorise the new language: Paschall's 'Botanick Tables'. Oxford, Bodleian Library, MS Ashmole 1820b, fol. 56r.

Far more aware of what was at stake were Ward and Pigot. Ward, it became apparent, had all along conceptualised the ideal artificial language as analogous to what he and his contemporaries termed 'specious Analytics' (Ward and Wilkins 1654: 21), or what is now called algebra. Hence his language sought to reduce reality to 'extreamely few' 'simple notions' 'by the helpe of Logick and Mathematiticks'. These notions would each be given a sign, and hence a language written down in this fashion would show its thought content in its very notation, much as an algebraic expression shows exactly how many *a*s and *b*s it contains, and is not merely an irreducible product. A complex algebraic expression is, at it were, its own anatomy; but the product of an arithmetical sum is simply a number. So while Wilkins pursued his classificatory tables, Ward was interested in the logical relation of ideas. Pigot, for one, saw clearly that Ward's artificial language was incompatible with Wilkins's venture, and in his opinion would prove too cumbersome to use. A 'logical' language threatened to collapse into overresolution, in which everyday notions were replaced with long, complex analytical conglomerations, and common concepts such as 'chair' or 'dog' would end up as sprawling compounds. In Hooke's well-chosen phrase, this was merely 'a remedy far worse then ye Disease'. Paschall eventually conceded that the revision group, although it had set out to revise and augment Wilkins, had in effect discovered a schism between possible models of an artificial language, and the problems with both alternatives. Lodwick and Paschall tinkering with their lists of augmentations threatened to turn the language into a classificatory Babel; whereas it was not obvious how Ward could turn his logically inclined proposals into a language that worked at all.

Ironically Wilkins was wise to stop developing his language where he did. His book, though explicitly only an essay 'towards' a finished product, actually works rather well within its limitations, and even today it is possible to write fairly complex text in the script, although one is of course imprisoned within the world-view within which Wilkins's own tables are themselves trapped. The revision group merely discovered that extensive reform, either within or of the basic Wilkinsian system, resulted in a solution worse than the initial problem. So the revisers failed, but they failed in an illuminating way. At the practical level, Aubrey realised that any workable artificial language would have to be absorbed into the school curriculum if it were to have any chance of social success. But how could this be achieved? In one of his private manuscripts, Lodwick envisaged what would need to happen. At some point in the mid-1670s he completed a draft of a utopian fiction in which an artificial language is imposed by a strong central government upon his entire utopian nation, with schools reformed according to its principles, and adults forced to acquire fluency, or forfeit citizenship, and ultimately suffer exile. But the more interesting failures were at the theoretical level. Aubrey had complained, explicitly with Wilkins's language in mind as the antidote, 'we make Rules from ye Language; whereas a Language should be made from Rules'. But a rule-bound 'philosophical' language can only be as good as the philosophy informing the language, whether based on logic or on classification, and as new ideas and things are being thought, made or discovered on a daily basis, an artificial language would have to be adaptable, and to have a mechanism for creating, absorbing and locating new words in its structure. Such a language, philosophically expressive but capable of evolving, becomes indistinguishable in practice from a complex natural language, and just as hard if not harder to learn. Some of Aubrey's more profound contemporaries realised this: John Wallis, for one, never denied the feasibility of constructing an artificial language, but he was unconvinced that such a language was *necessary*. Wallis regarded the languages of Dalgarno and Wilkins as interesting experiments, but he was not convinced that they told us anything philosophically more profound than the natural languages. The late-nineteenth- and early-twentieth-century artificial languages Esperanto, Ido and Interlingua have all tacitly acknowledged Wallis's scepticism: these are all derived from (European) natural languages, and offer grammatical simplicity and ease of acquisition, but no new insights about the nature of reality or how to organise it.

It is often said that the seventeenth-century artificial language movement was killed off by the coming of the philosophy of John Locke, whose *Essay Concerning Human Understanding* (1690) proposed a radically disruptive model of how humans acquire and associate sense-impressions. This is not quite true. As we have seen, the various language planners generated quite enough controversy among themselves to call the future of the project into question, and indeed it is only the length of involvement of Aubrey and Lodwick that really allow us to call this a 'movement' at all. By the late 1670s, as we have seen, it was increasingly obvious to those revising Wilkins that they faced insuperable problems. The 'movement', never all that coherent, was not killed, but ran out of steam. To be sure, it had a life on the continent in the eighteenth century, particularly due to the enthusiasm of Leibniz and the resurrection among some oriental scholars of the hypothesis that Chinese had originally been an artificial language, as Hooke had once argued, but in Britain the movement died. If this was not Locke's doing, it is nevertheless useful to ponder why his philosophy shut the door on any revival.

Wilkins and his contemporaries relied on certain shared assumptions about human psychology. Certainly, human thinking is interfered with by human passion, and this was largely identified as a consequence of the Fall of Man: unclear, indistinct or overpassionate thought is evidence that we are all blighted by original sin. Nevertheless, we can still grope after the ideal of cognition, we can still strive to have clear and distinct ideas, and we can trust that when our brains are working well, they are all working alike. My perception of the world is, from the point of view of the settled intellect,

*A VOYAGE TO LAPUTA, ETC.* 257

tage in point of health as well as brevity. For it is plain that every word we speak is in some degree a diminution of our lungs by corrosion, and consequently, contributes to the shortening of our lives. An expedient was therefore offered, that, since words are only names

for things, it would be more convenient for all men to carry about them such things as were necessary to express a particular business they are to discourse on. And this invention would certainly have taken place, to the great ease as well as health of the subject, if the women, in conjunction with the vulgar and illiterate, had not threatened to raise a rebellion, unless they might be allowed the liberty to speak with their tongues, after the manner of their forefathers; such constant irreconcilable enemies to science are the common people. However, many of the most learned and wise adhere to the new scheme of

2 K

**Figure 41** Wilkins parodied: Jonathan Swift's Lagadan philosophers from *Gulliver's Travels*, reduced to communicating by ostending objects, from an illustrated edition of 1886. Oxford, Bodleian Library, 256 d. 20, p. 257.

the same as yours. When we look at a tree, both of us apprehend in our mind that tree as a simple idea.

Locke turned this model of cognition upside down and, in so doing, exposed the deep philosophical conservatism of the artificial linguists. Locke argued that it is pointless to speak only of what we perceive clearly and distinctly, and that all people do not apprehend the same simple ideas when looking at the same objects. Humans do not at first perceive simple things 'out there' in the world, but complex, highly personalised agglomerations of sense-data, which we only gradually learn how to break down into the simpler elements out of which shared language is made. Rather than immediately perceiving the basic things out of which a Wilkinsian language is to be made, we only come to these notions quite late in the long process of learning how to think, and we all come to them by subtly, autobiographically different routes. Every tree to me is a different mixture of sounds, shapes, colours and smells. The artificial linguists had therefore got it the wrong way round. They were not holding up a mirror to cognition, but a falsely artificial and quite arbitrary imposition, with little relation to human intellection as it actually worked. Therefore, as Locke sealed the matter, no one can 'pretend the perfect *Reforming* the *Languages* of the world, no not so much as that of his own Country, without rendering himself ridiculous'. Locke was attacking simultaneously Aristotelian and Cartesian teachings about both logic and taxonomy, and it was on these very teachings that the language planners relied. The language planners protested their modernity and their desire to render philosophical, religious and social life better than it was. These were surely laudable aims. But Locke demonstrated that despite the obvious innovation of much of the scientific revolution, when it came to constructing a 'new' scientific language, the artificial linguists proved unable to resolve, or were unaware of, the contradictory currents flooding their efforts.

When retrospective ridicule came to the artificial linguists, it was in the unforgiving form of Jonathan Swift's *Gulliver's Travels* (1726). In the academy of Lagado, Gulliver finds men who, in a *reductio ad absurdum* of Wilkins, have decided to outlaw words and deal only with things. So they carry on their backs huge sacks of objects, which they brandish at one another instead of talking. It is a cruel judgment. To be sure, future artificial linguists learned not to make the philosophical claims that so excited Wilkins and his contemporaries. But it is pusillanimous to prefer the company of Swift to the bubbling enthusiasm of Lodwick, of Wilkins, of Ward, of Dalgarno, of Ray, and of Aubrey himself, the whetstone who made others sharp.

# 6

# Megaliths

Today, perhaps the most celebrated of Aubrey's contributions to the scholarship of his age was his study of ancient megaliths. His 'Monumenta Britannica', however, remained in manuscript – and is still substantially unedited – and so the core component in which he discussed such monuments, the 'Templa Druidum', was not easily available to his contemporaries. The manuscript, deposited in the Ashmolean Museum by 1693, but lent out again soon afterwards, was nonetheless consulted by prominent scholars and antiquaries in Aubrey's own time; and elements of his work, explicitly attributed to Aubrey, were printed in the massive 1695 revision of William Camden's *Britannia*, published two years before Aubrey's death, and edited by the young scholar Edmund Gibson. (Although Thomas Tanner, the probable reviser of the relevant section, rehearsed Aubrey's theory of British origin, he opted finally for the compromise that megaliths in general were British, but that Stonehenge in particular, because of its 'architraves', imitated aspects of Roman architecture.) In the next century Aubrey's manuscripts deeply influenced the researches of the antiquary and clergyman William Stukeley (1687–1765), who however coarsened Aubrey's analysis by overstating his case for Druidic religion as a kind of proto-Christian monotheism, a polemical twist quite alien to Aubrey's more tentative, gentle methodology. Although Aubrey decisively influenced his own contemporaries, his failure to publish in print meant that the true nature and extent of his influence has had to be rediscovered by modern historians and archaeologists, and Aubrey is honoured today through the 'Aubrey Holes', the circle of depressions surrounding Stonehenge that Aubrey may or may not have discovered. But once again it is perhaps best to use Aubrey as a lens onto the varied discussions of his age, rather than a lone innovator, and this is the task of the current chapter.

Speculation on megaliths, and on Stonehenge in particular, was an old activity by Aubrey's time, and had exercised Elizabethan antiquaries and poets. Field trips to Stonehenge had been undertaken by John Twyne in the earlier sixteenth century; and John Jewel, the famous Elizabethan ecclesiastical apologist and Bishop of Salisbury, proposed, as Inigo Jones would repropose in the early Stuart period, that the ruins were Roman in origin. But sheer puzzlement and awe were still respectable reactions. As Samuel Daniel (1562–1619) stated in his poetic defence of learning, *Musophilus* (1599),

And whereto serue that wondrous *trophei* now,
That on the godly plaine neare *Wilton* stands?
That huge domb heap, that cannot tel vs how,
Nor what, nor whence it is, nor with whose hands,
Nor for whose glory, it was set to shew
How much our pride mockes that of other lands?
Whereon when as the gazing passenger
Hath greedy lookt with admiration,
And faine would know his birth, and what he were,
How there erected, and how long agone:
Enquires and askes his fellow trauailer
What he hath heard and his opinion:
And he knowes nothing. (Daniel 1599: sg. C2v)

The chroniclers likewise were not sure what to do with the Wiltshire stones, garbling older stories of their magical passage from Ireland. As Shakespeare's major historical source Raphael Holinshead doubtfully mused,

> How and when these stones were brought thither, as yet I cannot read; howbeit it is most likelie, that they were raised there by the Britons, after the slaughter of their nobilitie at the deadlie banket, which Hengist [i.e. the fifth-century King of Kent] and his Saxons prouided for them, where they were also buried, and Uortigerne their king apprehended and led awaie as captiue. I haue heard that the like are to be séene in Ireland; but how true it is as yet I can not learne. The report goeth also, that these were brought from thence, but by what ship on the sea, and cariage by land, I thinke few men can safelie imagine. (Holinshead 1587: 129)

Such interpretations can be traced throughout the early seventeenth century. In 1627 the recusant scholar Edmund Bolton (1574/5–1634 or after) proposed that Stonehenge was the tomb of Boadicea, Queen of the Iceni, but all that is recalled of his effort now is his fine phrase, 'these orderly irregular, and formlesse uniforme heapes of massiue marble' (Bolton 1627: 182). The heyday in the earlier period for henge speculation proved to be the decade 1655–65, bounded at either end by the publications of the architect Inigo Jones and the physician Walter Charleton. Jones claimed that Stonehenge was a Roman temple of the Tuscan order; Charleton countered that it and similar megaliths were Danish coronation sites in origin. Jones was edited and supported by his disciple John Webb, architect and amateur sinologist; Charleton's collaborator and eventual corrector was John Aubrey himself.

Inigo Jones (1573–1652) was first set the task of examining Stonehenge by King James VI and I when the latter was on Progress in Wiltshire. Sent for by the Earl of Pembroke (based in nearby Wilton), Jones was asked for his opinion on the puzzling structure. Jones, so he claimed, had studied ancient monuments abroad, but Stonehenge impressed him

**Figure 42** Stonehenge, as imagined by Inigo Jones, based on four superimposed equilateral triangles inscribed in a circle, from *The Most Notable Antiquity of Great Britain vulgarly called Stone-Henge on Salisbury Plain Restored* (London, 1655). Oxford, Bodleian Library, C 2.25 Art.Seld, plate between pp. 60 & 61.

as unique, 'being different in *Forme* from all I had seen before: likewise, of as beautifull *Proportions*, as elegant in *Order*, and as stately in *Aspect*, as any' (Jones 1655: 1). Jones, who had held the office of Surveyor of the King's Works since 1615, is of course celebrated as the 'English Vitruvius', devoted to the theorists of Roman classical architecture, and largely responsible for their decisive influence upon English taste. In retrospect it is perhaps not surprising, therefore, that Jones talked himself into a belief that Stonehenge, too, with its 'beautifull *Proportions*', was a classical monument. His papers on the subject were posthumously edited by his disciple John Webb, and published as *The Most Notable Antiquity of Great Britain, vulgarly called Stone-Heng, on Salisbury Plain restored by Inigo Jones* (1655). The work as assembled by Webb opens with a rebuttal of the idea that Stonehenge could have been the product of an indigenous Druidic culture, producing an argument that would become a motif of such discussions. The Druids, as documented by Julius Caesar in his *Gallic Wars*, practised their rites in groves, and the culture over which they presided was not technologically advanced. Therefore they could not have raised such a monument as Stonehenge, and the only competent visitors to England capable of doing so were the Romans. This Jones confirmed by detecting among the remains of Stonehenge the traces of a mathematical order prescribed by Vitruvius, of Tuscan columns arranged following four superimposed equilateral triangles inscribed in a circle, the Vitruvian recommendation for the orchestra and stage of a Roman theatre. The result was an open-topped temple dedicated to the Roman God Coelus.

Jones's theories have subsequently met with much derision. But Jones introduced two important methods. First, he examined the object under study itself, and did not rely merely on written accounts; as Surveyor of the King's works, he indeed surveyed. Second, he drew the site, and Webb published many of these drawings in the final book. These included perspective illustrations of the stones as they now appeared, and also geometrical plans of how the inner circles now appeared, and of how Jones proposed they had originally been arranged. In other words, he used Vitruvian geometrical principles to reverse the ravages of time. Jones clearly over-idealised, as Aubrey was to point out, but he had taken the crucial step of bringing to the study of such ancient sites the skills of both the theoretical and the practical mathematician. Jones was fatally hampered by deference to textual sources and

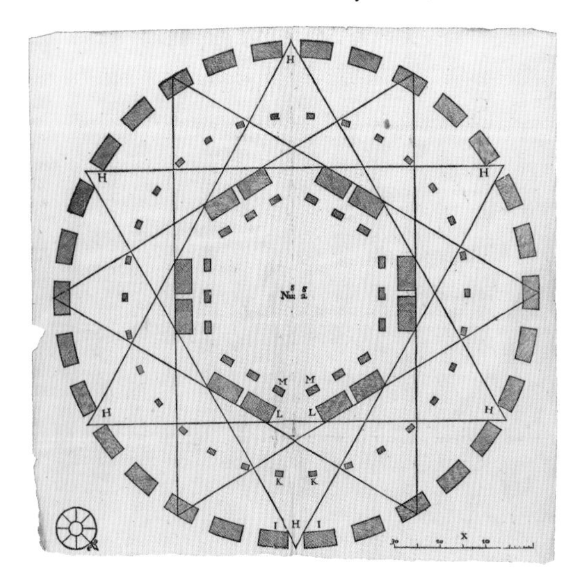

geometrical idealism, but, despite the error of his conclusion, his method was praiseworthy: he sought to harmonise the critical skills of the historian with the mathematical skills of the surveyor; 'I have now proved [my theories] from Authentick Authors, and the rules of Art [i.e. mathematics]', as his treatise concludes (1655: 107).

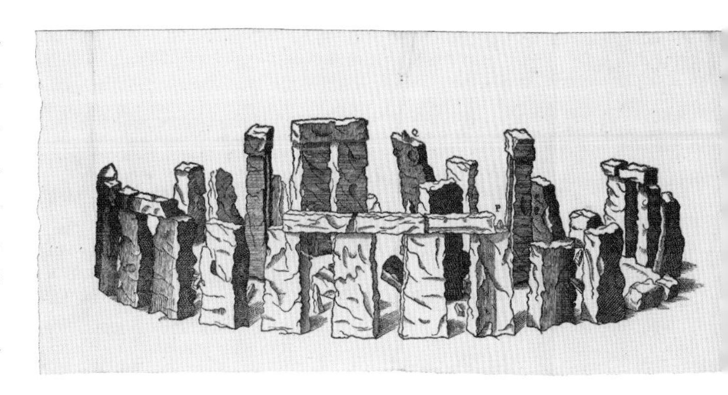

**Figure 43** Walter Charleton's less mathematically determined Stonehenge, from the *Chorea Gigantum* (London 1663). Oxford, Bodleian Library, A 2.19 Linc, plate between pp. 10 & 11.

Jones's book was soon answered by Aubrey's friend the FRS and physician Walter Charleton (1619–1707). Charleton had successfully proposed Aubrey to the Royal Society in January 1663; and by July he was presenting the Society with materials which it is reasonable to suppose he obtained from Aubrey himself:

> Dr. Charleton presented the society with the plan of the stone antiquity at Avebury, near Malborough, in Wiltshire; suggesting, that it was worth the while to dig there under a certain triangular stone, where he conceived would be found a monument of some Danish king. (Birch 1756–57: 1.272).

It seems likely that Aubrey also made a presentation at this meeting, and certainly Charleton's carefully drawn plan survives in the Royal Society's archives, as does Aubrey's own plan. Aubrey and his friend Sir James Long were therefore asked to investigate. But as Charleton's comment on the triangular stone (known today as the 'Cove') suggests, he had already decided upon a hypothesis differing from that of Jones. This he had just published in the same year under a title (and a title page) deliberately imitating and redirecting that of Jones: *Chorea Gigantum, or the Most Famous Antiquity of Great-Britain vulgarly called Stone-Heng, standing on Salisbury Plain, Restored to the Danes.*

Charleton's short book performed two swift operations, destructive and then constructive. Attacking Jones's textual arguments, he noted the complete silence of ancient historians, a stumbling block that had indeed required of Jones some special pleading. Attacking Jones's architectural arguments, Charleton accused him of selective quotation from Vitruvius, and of misrepresentation of the actual monuments. Where Jones wanted Roman columns, Charleton saw only '*unwrought*, *rough*, and *rude*' stones – Jones had attempted to evade this problem by claiming that the stones were ceremonially carved in the imitation of flames – and where Jones detected regularity, Charleton could comment that 'the most curious eye can scarcely find a perfect similitude in any two of them' (1663: 30). Instead, Charleton turned to a new kind of comparative archaeology, tracing continuity not from ancient classical practice, but from the more recent monuments of the northern European nations,

**Figure 44** Oxfordshire's Rollright Stones, as they appeared in Aubrey's time, from Robert Plot's *The Natural History of Oxford-Shire* (Oxford 1677). Oxford, Bodleian Library, Ashmole 1722, from tab. XVI.

in particular Denmark. In this he drew heavily on the work of the famous Danish scholar and collector Ole Worm ('Olaus Wormius', 1588–1655), whose *Danicorum monumentorum libri sex* (1643) 'furnish[ed] me with not only verbal Descriptions, but lively Draughts or Pictures also of sundry Antique *Danish* Monuments', as Charleton said (1663: 30). Indeed, Charleton's work was little more than an epitome of Worm applied to Stonehenge. Charleton was in one sense heading in exactly the wrong direction – he dated the construction of Stonehenge to the beginning of the reign of Alfred; that is, in the mid- to late ninth century. But in another sense he had practised as well as preached the principle of comparativism in archaeology, and this was a contribution of lasting importance. Granted, Charleton relied almost entirely on Worm's book, but at least it was a book of modern field research, carefully described, classified and illustrated. Unsurprisingly, Danish scholars were themselves flattered by Charleton's work, an early instance of Scandinavian–English archaeological exchanges that were to continue throughout the century. Charleton won some lasting converts among the English, too. When Robert Plot came to survey the issue in his *Natural History of Oxford-Shire* (1677), he agreed that the Oxfordshire equivalent, the Rollright Stones near Chipping Norton, were Danish or Norwegian in origin. But Plot also noted that Edward Stillingfleet had proposed that the Yorkshire and Wiltshire megaliths dated back to ancient British times, and were possibly the very stone idols the ancient Britons once worshipped; Stillingfleet, too, had appealed to Worm's Danish parallels. Plot hedged his bets in a prescient way: some stone structures, he decided, are only as old as the period of the Danish invasions; others are considerably more ancient (1677: 337–45; Stillingfleet 1676: 387).

Charleton's book was certainly not a disinterested piece of scholarship. It was dedicated to Charles II, whose visit to Stonehenge after his flight from the Battle of Worcester is recalled in Charleton's preface; and there too Charleton took occasion to play upon the Restoration of the king, and his own 'restoration' of Stonehenge. The work was prefaced by some verses, including a celebrated poem by John Dryden, also an (otherwise inactive) FRS, in which he lauded English science's successful attack on the philosophy of Aristotle, in the persons of Francis Bacon, William Gilbert, William Harvey and Robert Boyle. In other words, Charleton's book was packaged as an advert for the new science and the new Royal Society, claiming privileged access to the King himself, and embodying a new principle of field investigations: the comparison of objects with other similar objects, both at home and abroad. It is not surprising, therefore, that Charleton was so fiercely and intemperately attacked by Jones's disciple Webb in his reaffirmation of the Roman thesis, *A Vindication of Stone-Heng Restored* (1665). Webb, a marginalised figure, whose veneration for the achievements of the ancients would soon lead him to make some fascinatingly bizarre claims on behalf of ancient Chinese culture, clearly felt threatened by the new science; and the disputes over the origins of Stonehenge and Avebury must also be read as in part disputes about the proper sources of authority. The Jones–Webb–Charleton altercation was not just a quarrel between antiquaries, but an early engagement in the battle that came to dominate the juncture between the early-modern period and the enlightenment, the so-called quarrel of the 'Ancients' and the 'Moderns'.

If we were to consult only printed books, it would appear that Jones (posthumously) introduced mathematics to the emergent activity of archaeology, while Charleton contributed the idea of comparing like but distant objects. But such developments are not simply a matter of publication, and indeed John Aubrey's own field researches began before and concluded long after the decade separating Webb's two visits to the press. In January 1649, while hunting with friends, Aubrey famously 'discovered' Avebury:

> the chase led us (at length) through the Village of Aubury, into the closes there: where I was wonderfully surprized at the sight of those vast stones: of w$^{ch}$ I had never heard before: I observed in the Inclosures some segments of rude circles, made with those stones, whence I concluded, they had been in the old time complete.

Aubrey came across Webb's edition of Jones six years later, and although he 'read [it] with great delight', he gently complained that Jones had 'framed the monument to his own Hypothesis, which is much differing from the Thing it self'. Later still, after Aubrey and his friend James Long had spent many visits by turns hawking and surveying at Avebury, his researches came to the attention of King Charles II, who had been told about Avebury by Charleton himself, and the mathematician and first president of the Royal Society, Sir William Brouncker. Aubrey, so the King was informed, had claimed that Avebury 'did as much excell *Stoneheng*, as a Cathedral does a Parish church'. The king therefore summoned Aubrey, who presented him with a 'draught' of Avebury done by memory. Charles was impressed, and Aubrey in due course took the King on a personal tour of the site. This, then, was the real cue for Charleton's presentation on Avebury to the Royal Society. That September, Aubrey returned to Avebury, which he now surveyed more accurately by using a plane table, widening his tour to take in

**Figure 45** A plane table, by John Worgan, 1696. Oxford, Museum of the History of Science, MHS 18184.

Stonehenge too. Aubrey wrote up his survey of these and many other megalithic monuments over the next three years, and presented it to the King. But alas Aubrey failed to obey the King's order that he print his work, the original 'Templa Druidum', or 'Temples of the Druids' (MS Top. gen. c. 24; 1980: 17–22). As a result the vast, two-volume manuscript now in the Bodleian is a mixture of a recopied 'Templa Druidum' and later notes, the complete work now titled 'Monumenta Britannica' in imitation of Worm's *Monumenta Danica*. It has also been annotated by Aubrey's friends Thomas Gale and John Evelyn, as Aubrey himself proudly recorded. Aubrey certainly intended to publish this work: in 1666 he had his picture engraved for the frontispiece; and after a long period of dormancy he again tried to get the work printed in 1693. In 1695 he handed the whole manuscript over to his agreed publisher; but by his death two years later nothing had come of the book other than one 'forthcoming' advert. The manuscript itself remained in the publisher's until 1836, when it was acquired by the Bodleian Library.

Aubrey worked as a combination of an archaeologist and a mathematician. But, unlike Jones, he would not bend the former role to fit the presuppositions of the latter. As he wrote, he would turn the stones '[in]to a kind of Æquation: and (being but an ill Orator my selfe) ... make the Stones ... speake for themselves' (fol. 30r). His use of a plane table in surveying is particularly significant. This instrument was a flat board with a sighting alidade, onto which a piece of paper was fastened. Lines were then drawn directly onto the paper using the sighting alidade, and the instrument therefore functioned as a map-generator, run on geometrical principles. Aubrey was also a sophisticated historical analyst: noting that the ditch at Avebury lies *inside* the bank, he correctly argued that the whole site could not have been for military or defensive purposes, otherwise the ditch would have been on the outside. Aubrey was certainly not averse to some Danish provenances. After his first draft of the 'Monumenta' he compiled a manuscript of notes on and extracts from Charleton; and in his work on Surrey he proposed that labyrinths might be Danish in origin (MSS Aubrey 11; Aubrey 4, fol. 21ar). But in the 'Monumenta', noting that megaliths were found in remote areas never penetrated by the Romans or the Danish, Aubrey inferred that they were not built by visiting, temporary conquerors, but were indigenous and Celtic in origin. Perhaps the most

visually striking example of his method of making the stones 'speake for themselves' is the plan he drew of Stonehenge 'as it remains in the present year, 1666'. Opposite he redrew Inigo Jones's equivalent plan, to the same scale. Jones, we recall, thought that the vertices of three triangles inscribed in a circle designated the internal structure of the stones, and hence he produced a hexagonal arrangement of pairs of stones at the centre of Stonehenge. Aubrey's own, facing plan instead detects five more or less present pairs of stones, and uses their arrangement to conjecture a missing further two sets. Aubrey's internal arrangement is therefore heptagonal, 'not so beautiful' as a hexagon, and not at all a classical construction. With admirable restraint, he commented: 'Why might not then the seaven-sided figure in the foregoing scheme be made in relation to the seaven Planets, & seven daies of the Weeke? I can not determine: I only suggest' (Aubrey 1980: 82). Aubrey also extended his survey to the whole of Britain: after Avebury and Stonehenge, he turned to stone circles in Cornwall, Dorset, Oxfordshire, Cumbria, Westmoreland, Caernarvonshire, Denbighshire, then to Irish and Scottish counties, embracing not just stones but barrows, mounds, ditches and anything else that we might term prehistoric. He also adduced parallels from other cultures and countries. The 'Monumenta' became increasingly inclusive as Aubrey moved from his reconstruction of Druidic culture through to more historically recent topics including coin hoards, church architecture, even changing styles of shields, windows, clothing and handwriting, as we shall see in another chapter. By its conclusion the 'Monumenta Britannica' manuscript had broken not just the bounds of the original 'Templa Druidum' core, but also its own initially supposed perimeter, and in doing so transcended Worm's Danish model.

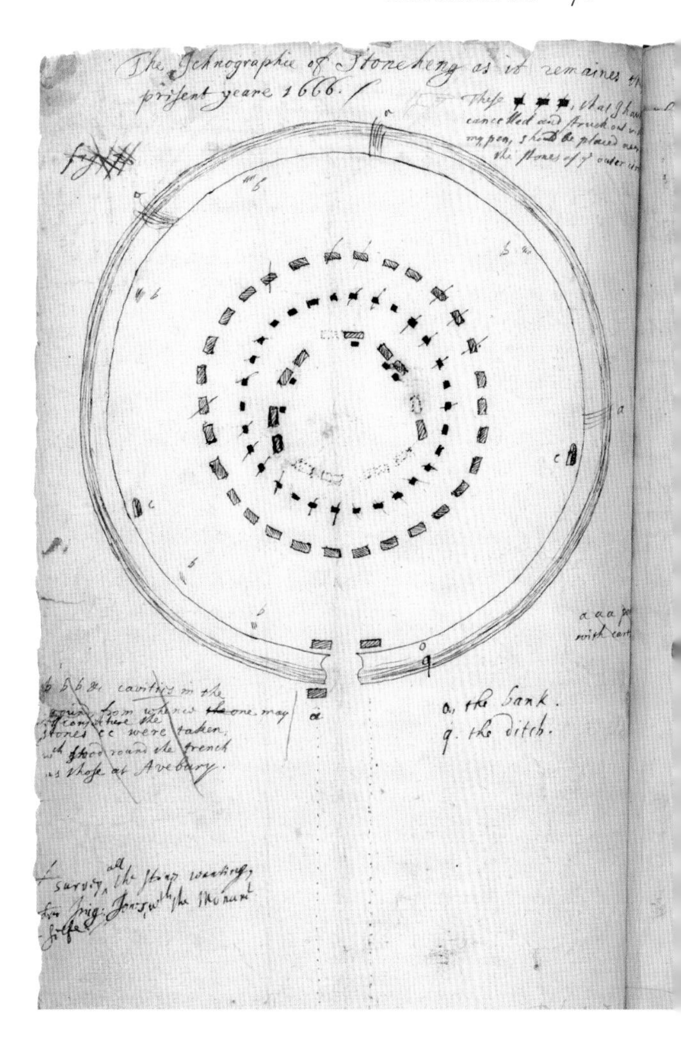


**Figure 46** Aubrey's heptagonal Stonehenge, drawn from first-hand surveying. Oxford, Bodleian Library, MS top. gen. c 24, fol. 64v.

Aubrey discussed his ideas widely and circulated his papers among friends, and the 1690s became the second great decade of henge speculation. 'I would willingly print my Templa Druidum in my life time; for that is finished; only wants an Aristarchus to polish the style', he wrote to Wood in mid-1690 (MS Wood F 39, fol. 405r). The philosopher John

Locke even offered to fund the publishing of the 'Monumenta'. Although all these plans came to nothing in Aubrey's lifetime, his manuscript was excerpted in the revised 1695 Camden. At first Aubrey was a little nervous that his hard work would be stolen from him, but he was accredited in the printed text. This, and a partial transcript made by the philologist and FRS Thomas Gale (1635/6–1702) and discussed below, kept Aubrey's ideas alive for a significant few in the early eighteenth century. Perhaps the most important new man on the scene was the second Keeper of the Ashmolean Museum, the Welsh scholar Edward Lhuyd, who as we have seen became a particular friend of Aubrey. Lhuyd was involved in many areas of scholarship, and was a contributor to the revised Camden. In Aubrey's final years Lhuyd became his closest Oxford contact, and he absorbed and carried forward Aubrey's thesis that the British megaliths were pre-Roman, although Lhuyd was careful not to ascribe one purpose or date to all such monuments – 'for my part I leave every man to his conjecture' (Camden 1695: 630). Even before the publication of the revised Camden, Aubrey was becoming known as the partisan of the 'British' theory. In an exchange in early 1694 which Lhuyd had with the brash John Woodward, Gresham Professor of Medicine and FRS, the latter wrote:

> You are certainly right in y$^{or}$ Opin. abt y$^{e}$ heaps of Stones; they were undoubtedly funeral Monuments. Role-right was such: Stone-henge on Salisb. Plain no other. M$^{r}$ Aubrey, & some others, I know, talk of British Temples, Druids temples &c but these Gentlemen indulge too much in fancy, wherever records are wanting, or rather ag$^{st}$ records, ag$^{st}$ reason & Antiquity.

Woodward instead proposed the old theory of Scandinavian origins, and in an interesting piece of bias for a man who was himself an avid field-worker and collector of fossils, he criticised Aubrey for not being sufficiently book-bound:

> As for M$^{r}$ Aubry, he may well be for ought I know a very worthy person; but, after this I think I may pretty freely pronounce him mistaken, if he contend y$^{t}$ Stonehenge was a British or Druids Temple: 'twere to be wisht y$^{e}$ writers on these subjects would not rely so much on fancy & conjecture, but consult y$^{e}$ Ancients, as those who are y$^{e}$ most proper Judges of these Matters. (MS Ashmole 1817, fols 367r–68v, 367v)

In the early eighteenth century, indecision persisted. The first full tract on Avebury to reach the press was the short *Avebury in Wiltshire, The Remains of a Roman Work* (1723), the argument of which is stated in the title. Written by Thomas Twining (1686–1739), vicar of Wilsford and also nearby Charleton, parishes just south of Avebury itself, the work consisted of the usual farrago of quotations from the Roman historians, stretched rather unconvincingly over the remains visible to Twining. Twining, who had studied at All Souls College in Oxford throughout the 1680s, merely proved himself to be almost entirely unaware of contemporary scholarship on the question, and although he drew on personal experience to comment sadly on how the stones had been the 'common Subject of Plunder' for time immemorial, he made little effort to chart the remains himself (1723: 3). Indeed, he used the depredations of the stones as an excuse to avoid doing so, instead providing a fold-out map of what his (solely textual) researches convinced him had been the original arrangement.

Yet the Aubreyan interpretation of deep antiquity was gaining some decisive ground, and with it the necessary corollary that the first makers of Stonehenge were pagan Britons. Indeed, support came from the very far north, in the form of Martin Martin's *Description of the Western Islands of Scotland* (1703), the research for which had been partially funded by subscriptions from the London virtuosi collected by Hans Sloane. Martin (d. 1718), a Skianach by birth, described – and included a plate of – the Callanish standing stones of Lewis, one of the Outer Hebrides, as well as standing stones elsewhere in the Hebrides. The Callanish stones formed a long avenue leading up to a circle, and when Martin asked the locals about their origins, he was told 'it was a place appointed for Worship in the time of Heathenism, and that the Chief *Druid* or Priest stood near the big Stone in the center, from whence he address'd himself to the People that surrounded him' (1703: 9). In Oxford, the librarian, antiquarian and scholarly editor Thomas Hearne confessed in the year of Twining's publication that he had once inclined to the Roman interpretation of Stonehenge, but had since gone over to the 'British' school, 'tho' I cannot account for the Methods they used in erecting it'. Two years later Hearne was to publish a transcript of the anonymous manuscript *A* 'Fool's Bolt Soon Shott at Stonage', in which it was suggested 'that Stonage was an old British triumphall tropicall temple, erected to *Anaraith*, their Godess of victory, in a bloudy field there, wone, by illustrious *Stanengs* and his *Cangick Giants*, from K. *Divitiacus* and his *Belgæ*' (Hearne 1725: 484). (Aubrey may have encountered this manuscript, as we know Paschall owned a copy.) A year previously, too, Hearne recorded that he had dined with the astronomer Edmond Halley, who 'hath a strange, odd Notion that Stonehenge is as old, at least almost as old, as Noah's Floud' (Hearne 1906–21: 7.89–90, 350). Halley had visited Stonehenge in 1720, and brought back a piece of it to Royal Society, where it was examined under the microscope by William Stukeley.

The FRS, antiquary and natural philosopher William Stukeley (1687–1765) is the most notorious of all megalithic speculators, and

The manner of laying on the impost at Stonehenge

**Figure** 47 William Stukeley's imagined construction of Stonehenge: 'The manner of laying on the impost'. Oxford, Bodleian Library, Gough Maps 231, fol. 11r.

Aubrey's own reputation has to an extent been tied to that of Stukeley. Before he was thirty he had made a model of Stonehenge, and he started his extensive series of field trips to Avebury and Stonehenge two years later. Not only did Stukeley generate a huge quantity of sketches, measurements and notes, but he collected previous materials on such monuments. Stukeley was deeply indebted to Aubrey's work, which he accessed through notes from it made by Thomas Gale. Stukeley's copy of the 1695 *Britannia* also survives in the Bodleian today, in which he has underlined Aubrey's Druidic remarks. He profited greatly, too, from Lhuyd's unpublished illustrated reports on his own surveying expeditions. Stukeley's own manuscript treatise 'The History of the Temples of the Ancient Celts' is generous in its citation of Aubrey and Lhuyd, but Stukeley was very slow to acknowledge his debt to Aubrey in any printed work, and so far as reception is concerned Stukeley veers close to plagiarism, and he certainly diminished the reputations due to both Aubrey and Lhuyd (Bodleian Library, MS Eng. misc. c. 323). His labours came to fruition in two celebrated books, one on Stonehenge in 1740, and a further volume on Avebury ('Abury') three years later. The permanent value of Stukeley's work as a record of since-vanished features of the megaliths is undisputed. But Stukeley had always been fascinated by Druidism and the nature of ancient British religious belief, and he projected onto the megaliths a Druidic religion that was very close to the rational, 'natural' religion that had been promulgated by Noah and that underpinned, and reached its fullest expression in, Christianity. It is due to Stukeley that henges and Druids have become intertwined in the modern imagination, despite the chronological (and theological) fallacy on which the association rests. As a result, history has been rather ambivalent about Stukeley. Aubrey had certainly claimed that the megaliths were Druidic, but he had not pushed his interpretation, and he had certainly not turned his Druids into gentle proto-Christians.

Nevertheless, Stukeley's philosophical views, although unusual, are far from incomprehensible in context. Indeed, many of the FRSs of Aubrey's seventeenth century would not have found his practices uncongenial. Several of the founding fellows, including John Wilkins and Seth Ward, had been instrumental in developing a 'latitudinarian' theology which stressed the reasonableness of Christianity, and the basic availability of its core truths to any correctly functioning mind. And as all cultures were thought to descend from Noah via the Confusion of Tongues at Babel, arguments about primitive religion were really just arguments about how quickly, or even if, the original wandering tribes had forgotten their Noachic philosophy. Stukeley was historically mistaken, but his was an intelligible interpretation.

Even if Stukeley's religious thought is not as cranky as is often assumed, his significance lies in his development of the historian's use of physical evidence. It is significant that Stukeley appealed to Halley's reasoning for the antiquity of Stonehenge directly from the physical appearance of the monuments themselves, estimating the age of Stonehenge – '2 or 3000 years old' – by appraising the weathering of the stones. Stukeley also appealed to Halley's work in the 1690s on the variation of the magnetic poles in order to calculate when the magnetic compass-bearing architects of Avebury and Stonehenge had begun their work. Measuring the current alignments of the megalithic complexes and then back-extrapolating from Halley's proposed 700-year cycle, Stukeley came up with the construction dates of 460 BC for the (also less weathered) Stonehenge, with Avebury dating from perhaps 700 years before then. Stukeley, it is true, yoked his 'observational' hypotheses to his vision of a pure Druidic religion that has much in common with Isaac Newton's own contention that Stonehenge was a piece of Noachic symbolism representing the heliocentric system. But his practical methodology was 'experimental', unlike Newton's purely speculative, textual approach (Stukeley 1740: 5, 57; 1743: 52–3). This was the real legacy of Stukeley, and, behind him, of Jones in the 1620s, of Charleton in the 1650s, of Lhuyd in the 1690s, and of Aubrey right across the decades from the 1640s until his death fifty years later. Late last century, Aubrey himself was even claimed retrospectively as a Druid, who apparently in around 1694 revived a society of Druids originally created in Oxford in 1245, a circle called the Mount Haemus Grove. So he and John Toland the Deist – with whom Aubrey indeed discussed megaliths – sombrely donned their robes and performed Druidic rites. It is safer to dwell upon Aubrey's real achievements.

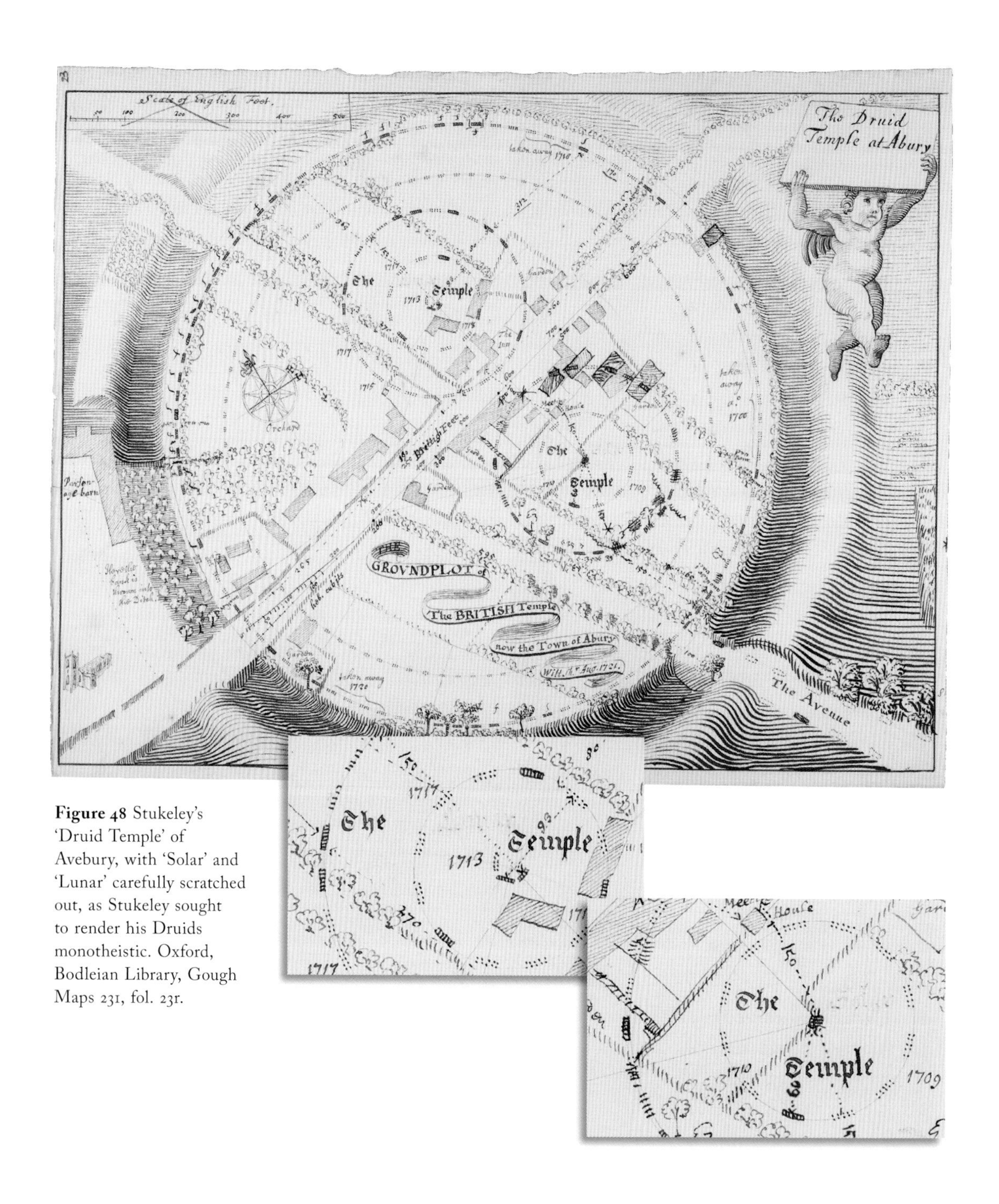


**Figure 48** Stukeley's 'Druid Temple' of Avebury, with 'Solar' and 'Lunar' carefully scratched out, as Stukeley sought to render his Druids monotheistic. Oxford, Bodleian Library, Gough Maps 231, fol. 23r.

# 7

# Aubrey and the Earth Sciences

The early-modern study of megaliths can most easily be described as an offshoot of chorography, or the study of specific locales. Yet, as we saw, the development of hypotheses about the origin, construction and purpose of megaliths was swiftly attracting, on the one side, questions about population dispersal and world chronology, and, on the other, questions about the stones themselves, their position, substance and form. This latter set of questions borders on the study of what we term today geology or more broadly the earth sciences. 'Geology' as a word only properly entered English in the eighteenth century; but as a concept it was solidifying in Aubrey's time, not least because of a number of widely discussed disputes about how to relate issues of geology (the study of the earth) and geomorphology (the study of the changes in the earth over time) to conventional biblical ideas about the age of the world and its supposed history. This set of debates, and Aubrey's contributions to them, are the subjects of the current chapter.

Aubrey began to write his *Natural History of Wiltshire* in earnest in 1656, and continued to work on the manuscript intermittently for three and a half decades. As a genre, the county history was an Elizabethan development, but this kind of antiquarianism was usually restricted to what we would think of as human geography – the relations of communities to their surrounds, with an emphasis on the built environment (i.e. castles, houses and churches, including their inscriptions), and on local family history (pedigrees, heraldry). This kind of geography might include discussions of famous earthquakes or notable springs, but the focus was still on settlers and settlements.

The seventeenth century saw a number of innovations supplementing, and eventually transforming, this basic model. Systematic experimentation on the environment as laid out in detail in Francis Bacon's *Sylva Sylvarum* (1627) encouraged organised attention to the physical properties of locales, ranging from soil content to acoustic idiosyncrasies. Again, the rise of Paracelsian medicine, with its attendant interest in 'iatrochemistry', or the treatment of disease with chemical substances, encouraged renewed attention to naturally occurring chemical remedies, especially in springs and spas. Coins and inscriptions had been a staple of county history since the Elizabethans, but more space was now being given to 'mute' objects, such as pots or bricks, and their substance and manner of construction. Aubrey's adjective 'natural' in his *Natural History* therefore signals that at least

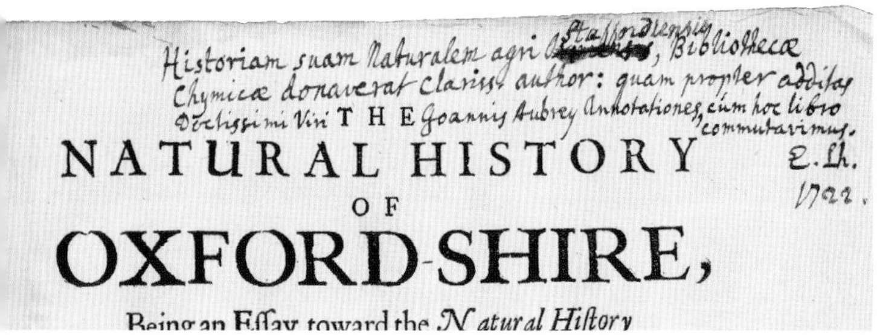

Historiam suam Naturalem agri Staffordiensis, Bibliothecae Chymicae donaverat Clariss. author: quam propter additas Doctissimi Viri Joannis Aubrey Annotationes cum hoc libro commutavimus. E. Ll. 1722.

THE
NATURAL HISTORY
OF
OXFORD-SHIRE,
Being an Essay toward the *Natural History*

**Figure 49** Aubrey's annotated copy of Plot's *Natural History of Oxford-Shire* (Oxford 1677), presented to the Ashmolean's 'Chemical Library', and exchanged for Plot's own initial presentation copy, as Edward Lhuyd's note explains. Oxford, Bodleian Library, Ashmole 1722, inscription on title page.

in theory he too would concentrate on physical rather than human geography.

Indeed, Aubrey's own verbal contributions to Royal Society debates frequently concerned topics that we would now group together under the earth sciences. He was interested in collecting meteorological data on winds, tides and barometric levels, or in arranging for it to be collected; he presented samples of iron ore and marl; he discussed pyrites springs; he reported a Rutland earthquake; and he was always interested in the commercial possibilities of natural resources, especially his native Wiltshire clays. For instance, on the penultimate day of 1675, Aubrey read out some of his observations on Wiltshire, and was asked as a result to obtain a type of soft Wiltshire iron ore, as well as some of the ultramarine clay from his own estate, Easton-Piers, 'which clay Mr. Doight supposed to be very fit for porcelane' (Birch 1756–57: 3.271). This last comment in the minutes is especially interesting, for the 'Doight' referred to is John Dwight (1633/6–1703), chemist and potter, who in the 1650s had experimented with Boyle and Hooke in Oxford, and who after the Restoration established a pottery at Fulham, quite frequently encountering Hooke and his circle in the coffee houses. Because Dwight had first developed his methods in Oxford in the interregnum, he was hailed eulogistically by Robert Plot in his *Natural History of Oxford-Shire* (1677) as the English rediscoverer of the foreign trade secrets of how to make Cologne ware, Hessian ware and China ware. This was particularly important for Plot, as Dwight's Hessian beakers and retorts were essential for chemical experiments, and Plot probably used Dwight's ware for his own work in the Ashmolean Museum (Plot 1677: 250–51).

Aubrey's contributions in these areas were generally well received by the early Royal Society. Parts of both his Wiltshire and his Surrey collections were read out, and in 1691 the Society arranged to have *The Natural History of Wiltshire* copied and placed in the Society's own library, where it is now MS 92. This largely scribal copy, which differs significantly from the manuscript now in the Bodleian, was a major commission, and an honour for Aubrey. The Royal Society clearly considered that Aubrey's style of natural history was a good synthesis of the various new elements entering into the compiling of regional histories; and, as we shall see, Aubrey also used the excuse of his natural county history to digress rather widely, and to lend support to the controversial geological theories of Robert Hooke. It is this larger context that concerns us now.

One chapter of Aubrey's *Wiltshire* concerns 'formed stones', or what we would call fossils. Aubrey considered that many of his Wiltshire fossils were the petrified remains of once-living things. This view – the view we now know to be correct – was nonetheless a controversial hypothesis in Aubrey's day. Part of the problem arose from the difficulty of classifying what was a fossil and what was not: in its broadest form, an early-modern 'fossil' could be anything from an ore to a mineral to chalk. The class of stone object in the form of something that looked like it once might have been alive was not an immediately separable category, because 'formed stones' included items possessing geometrical as well as potentially organic order, and the 'true' line between a pyrite aggregate and a coiled ammonite is not necessarily visible to a culture that does not possess our modern ideas of evolution and extinction. Nevertheless, natural processes of petrifaction were of the greatest interest to the early experimentalists because they showed that nature, like a Great Chemist, could seemingly transmute one substance into another.

There was little controversy over petrified trees dug from deep under the ground, and the ability of mineral-rich streams to coat or perhaps to transform bodies left to steep in them attracted wide comment too. The difficulty arose when dealing with objects that looked like they had once been fauna. Here there were a number of problems. First, many such fossils did not resemble recorded species, and this made naturalists very uncomfortable, as it was assumed that God would not allow any species to perish utterly, otherwise it would have been a mistake for Him to have created the species in the first place. Second, fossils were found in strange places – petrified sea shells at the top of hills nowhere near the coast, for instance. Again, many different forms in the same place appeared to be made of the same stone; whereas many similar forms in different places differed also in stone type. Why would a petrified sea urchin (they termed them 'brain-stones') in one place be made of a different substance than one found elsewhere? This was not how most living fauna behaved. Finally, there were adequate available explanations for fossils. In Aristotelian terms, they merely possessed the 'form' but not the 'substance' of living things, and nature might occasionally cross-materialise, as it were. Others pointed to the imitativeness and playfulness of nature, arguing indeed that it would be surprising *not* to find such 'correspondences' now and then. These arguments were often mystical or hermetic in inspiration, and an extreme example is the Jesuit Athanasius Kircher's *Mundus Subterraneus* (1665), a book eagerly awaited by the English experimentalists, even though most of

**Figure 50** Early chemical vessels excavated in 1999 from the foundations of the Old Ashmolean Museum, presently the Museum of the History of Science, possibly dating back as far as the period of the potter John Dwight's activity. Oxford, Ashmolean Museum, on permanent display in the Museum of the History of Science.

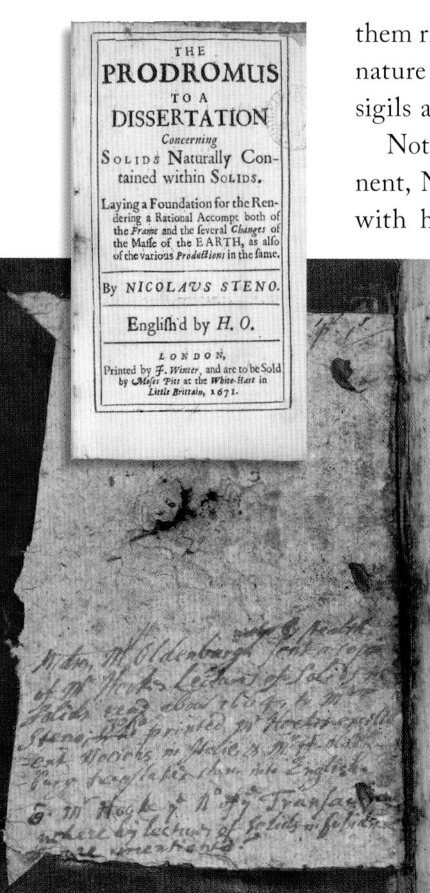

THE
PRODROMUS
TO A
DISSERTATION
*Concerning*
SOLIDS Naturally Contained within SOLIDS.

Laying a Foundation for the Rendering a Rational Accompt both of the *Frame* and the ſeveral *Changes* of the Maſſe of the EARTH, as alſo of the various *Productions* in the ſame.

By NICOLAVS STENO.

Engliſh'd by *H. O.*

*LONDON*,
Printed by *J. Winter*, and are to be Sold by *Moſes Pitt* at the *White-Hart* in *Little Brittain*, 1671.

**Figure 51** *Top* The founding text of stratigraphy, Nicolaus Steno's *Prodromus* (Florence 1659), in the translation of Henry Oldenburg (London 1671). Oxford, Bodleian Library, Ashmole C10, title page.

*Above* **Figure 52** Aubrey scrawls on the front board of his copy of Steno the accusation that the translator, Oldenburg, 'by stealth sent a copy of Mr Hooke's Lectures of Solids in Solids read about 1664, to Mr Steno, who printed Mr Hooke's excellent Notions in Italy and Mr H. Oldenburg translated them into English.' Oxford, Bodleian Library, Ashmole C10, front board.

them rightly regarded Kircher as credulous. Kircher purported to detect nature playing to the extent of painting mystical scripts and hermetic sigils all over unsuspecting minerals.

Not all, however, were content with these answers. On the continent, Nicolaus Steno effectively invented the discipline of stratigraphy with his *Prodromus* of 1669, in which he showed how the history of a landscape can be 'read' from its current appearance; and how fossils obviously differed in their origins from gems and crystals. Steno's work was translated by Henry Oldenburg in 1671, and Robert Hooke took the occasion to accuse Oldenburg of having betrayed Hooke's own ideas on the subject to Steno. (The accusation is unfounded, but Aubrey has faithfully recorded it on the inside board of his own copy of Oldenburg's Steno.) Hooke, indeed, was the most prominent English student of fossils. He vigorously defended their organic origin in his *Micrographia* (1665), a work so popular that its incidental analysis of fossil forms was bound to attract wide attention. Comparison with the minutes for Royal Society meetings at the time shows that Hooke was developing his theories against the background of discussions with several other fellows, but Hooke was to develop his theories far beyond the initial parameters set in such meetings. Hooke started with petrified wood, but soon extended his discussion to petrified shells, remarking that such 'Metamorphoses' can happen all over the vegetable and animal kingdoms. His analysis through the microscope of 'Serpentine stones' (i.e. ammonites) convinced him that they were the results of a petrifying liquor seeping into the original organic form, saturating its pores, and then solidifying, persisting after the original matter had decayed and perished. By displaying his conclusions as based entirely on observation and unaffected by theory, Hooke was attempting to present as uncontroversial a conclusion which was in fact highly problematic. Hooke knew this, and his intention was to persuade his readers that the modern appeal to scientific instruments was an intrinsically more reliable method than recourse to some prior, untested theory.

Needless to say, Hooke's work did not go unattacked. Indeed, his chief opponents came from within the Royal Society itself. The famous naturalist John Ray was willing to go with Hooke a certain way, but he saw and feared the problems caused by fossils of unidentified creatures found in seemingly impossible places. More bluntly, Martin Lister, a prominent physician and naturalist based in York, rejected Hooke's theories in a letter to the *Philosophical Transactions* in 1671, published by Hooke's enemy Oldenburg a mere few weeks before Lister himself was elected to the Society. In 1677 Lister was seconded by Robert Plot in his *Natural History of Oxford-Shire*. Lister was to be a major benefactor of Plot's Ashmolean. These men may have been able to accept the odd petrified shell near the seashore, but the weird stone creatures found by Hooke in inland quarries were to be classed as only ever stone in form, and to be sharply dissociated from genuine forms of life. This particular controversy raged throughout the 1670s, abated somewhat in the 1680s, and then reactivated in the 1690s, for reasons we will investigate.

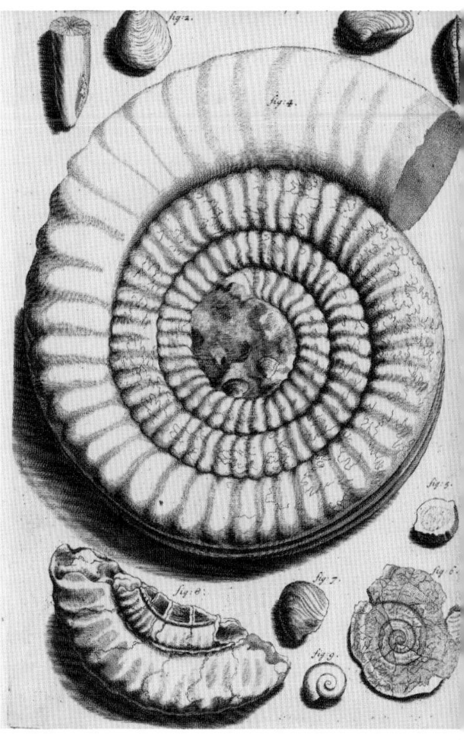

**Figure 53** Hooke's illustrations of fossils, including a large ammonite, from *Posthumous Works* (London 1705). Oxford, Bodleian Library, Savile LL 1, Table 6.

What was at stake here? One obvious problem with Hooke's organic fossils was that they implied that either some very strange animals had once lived in places where no trace of them can now be found, or that many fossils represented extinct creatures. Particularly this latter problem was felt to have a religious significance that persuaded most to err with Lister and Plot, rather than to explore with Hooke and (to some extent) Ray. Why would God create things imperfect enough to warrant extinction? A further, in the event far more far-reaching, problem was how to account for the positions of marine-type fossils in inland locations. Now the former problem could at least be met by Hooke by trumping the providentialist argument, which he did: if it was questionable for God to create things with only a finite usefulness in creation, then surely it was even worse to imagine Him creating *totally* useless objects. Hooke's rejoinder was a good one, and Plot could only counter it with some mumblings about the possible pharmaceutical uses of fossils. Hooke then went on to accept both alteration within and the extinction of given species, a development some modern commentators have wrongly hailed as a theory of evolution, but a remarkable position all the same. But it was the problem of position that really ignited Hooke's imagination, and Aubrey was one of the very few men who were willing to follow Hooke all the way down the path he took.

Hooke, in the long series of lectures he devoted to the subject from 1668 almost until his death three and a half decades later, set himself to explaining why marine fossils can be found inland. His answer was that land and sea are themselves in a constant if slow flux, and that the geomorphology of the earth tracks the wandering poles. As the poles of the earth drift around the continents over time, Hooke theorised, so the earth's surface fluxes in their wake, bulging out at the shifting equator. (He was here borrowing, via Edmond Halley, from Newton's demonstration in the *Principia* (1687) that a spinning globe will fatten slightly at its equator.) Hooke had therefore 'discovered' the cause of his fossils: the wandering poles cause slow vertical distending at ninety degrees, converting seas into mountains and vice versa, and hence over vast tracts of time all lands have changed places, and fish, as Hooke says, once swam over England.

Hooke's theories were shocking, and it is unsurprising that he never published these lectures. He faced serious problems of evidence: if land and sea swap places over time so completely as Hooke supposed, why have we so little record of it? The earth is after all only the few thousand years old as the Old Testament in all its variant versions suggests, and the Bible records only one such catastrophe in all history: the Flood. And what evidence is there that the poles have wandered so extensively? Hooke did not really manage to meet the latter objection, but to the former he proposed various answers over the years, including appeals to pagan mythology and poetry to supply what the Bible could not. The remainder of his lectures need not concern us, but it will be understood that Hooke's great challenge was one of chronology. If fossils were once organic beings, then the world really ought to be much, much older than the Bible assumes. Hooke and Halley eventually solved this problem by placing most of the earth's physical history in the 'Chaos' out of which the (current) world was (eventually) created, an ingenious piece of exegesis that again could not immediately be published at the time.

The next phase of speculation in the nascent earth sciences addressed precisely the biblical question. From late in 1680, the ecclesiastic and headmaster Thomas Burnet began to publish his *Telluris theoria sacra* ('Sacred Theory of the Earth', soon translated by Burnet himself into English as just *The Theory of the Earth*). This appeared in instalments up to 1689, and offered in stylish if rather prolix prose a biblical rereading of the earth's physical history from Creation to Conflagration. (Aubrey donated his copy of the English translation of the first two books to the Ashmolean; the marks in it are probably his own.) Burnet proposed that the earth had been created as a hollow, water-filled oval; and that, at the culmination of the excessive wickedness of the first generations of men, it had been providentially cracked open, releasing the waters within, which then formed the (otherwise physically impossible) Flood. Burnet, who owed far more than he cared or dared to admit to the recent physics of Descartes, was attempting to ground the Frenchman's widely studied system of geohistory upon a more rigorously scriptural basis. The book was a best-seller, as (usually horrified) scholars read how a churchman was using mathematics and experimental philosophy to rewrite sacred history. For Burnet's view of the antediluvian egg was so utterly unbiblical that he was forced to argue that Moses, supposed author of the Pentateuch,

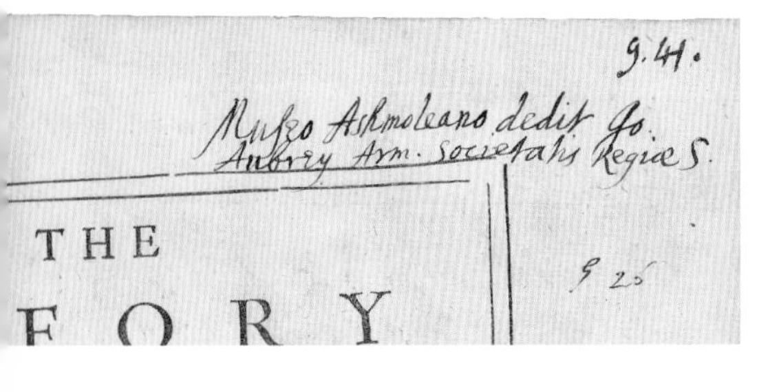


*Below* **Figure 54** Aubrey's inscription on the presentation copy of the English edition of the first two books of Thomas Burnet, *The Theory of the Earth* (London 1684). Oxford, Bodleian Library, Ashmole G 26, title page.

*Top right* **Figure 55** Edward Lhuyd's ninth class of fossils, 'Ichthyodontes Cuspidati', from his *Lithophilacii Britannica ichnographia* (London, 1699). Oxford, Bodleian Library, 8° K 60 Med, Table 15.

*Middle right* **Figure 56** Lhuyd's *Ichnographia* required sponsorship to meet the publication costs of an edition of only 120 copies: among those who financed the catalogue were Isaac Newton, and the naturalists Martin Lister, Tancred Robinson and Hans Sloane. Oxford, Bodleian Library, 8° K 60 Med, detail from sg. A1v.

*Bottom right* **Figure 57** Lhuyd's fossil 1260, a shark's tooth from Sheppey, Kent, 'where it is known to boys and girls as a *crampstone*'. This is the very fossil from which Lhuyd's plate derives. Oxford, University Museum of Natural History.

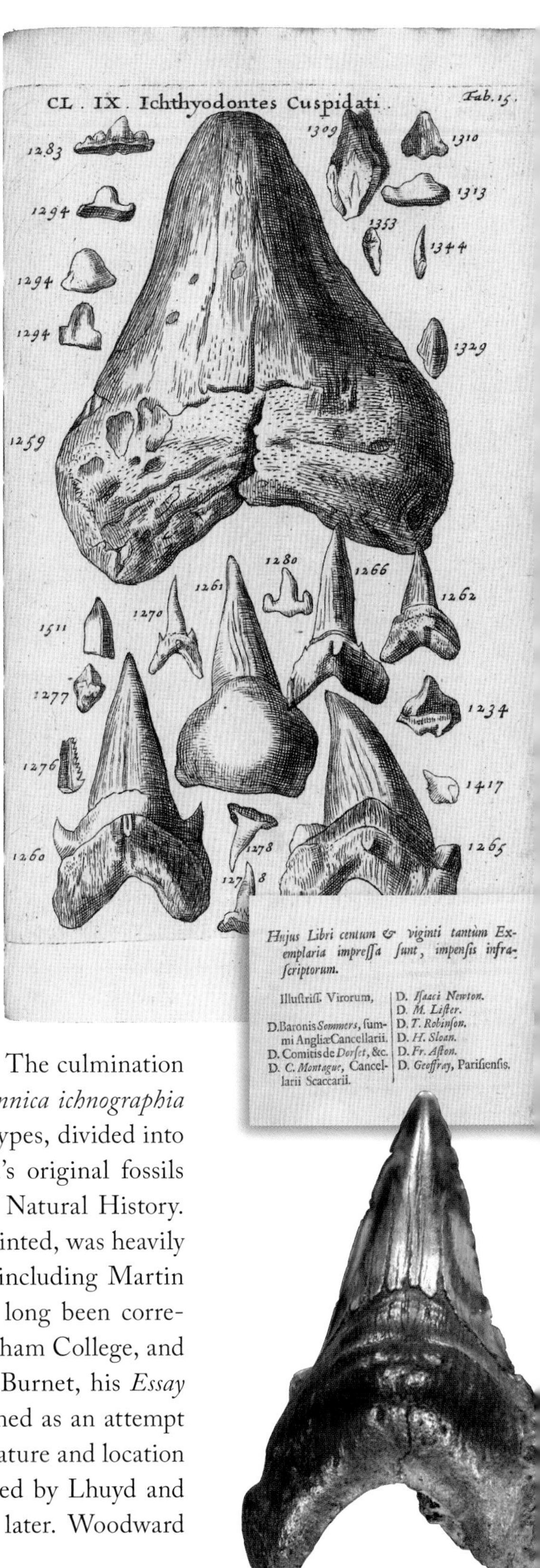


was writing only morally edifying material when he spoke of Creation, and was not at all concerned with scientific accuracy. Burnet's book, as most of his readers pointed out, was dynamite, for it implied that the new philosophy might no longer have to be shackled to the apparent literal meaning of scripture. This was an ironic turn of events for Burnet, who was presumably sincere in his desire to Christianise Cartesianism. Burnet had a few supporters, but they were very much in the minority, and usually of a Deistical complexion. Churchman Burnet certainly cannot have welcomed this radical appropriation of his theories, but it is hard to be fully sympathetic towards his gorgeously written but rather intellectually unstable book. He had started out trying to solve a biblical problem using the new philosophy. He had ended up demonstrating the dangers of such a project, and that made him a wanted man from either side's point of view. As a result, refutation of Burnet became a staple activity for both theologians and natural philosophers throughout the rest of the century and beyond.

Burnet did geomorphology on a grand scale, even down to the millennialist reconstruction of the earth after the Conflagration. But one area in which Burnet had been completely silent was that of fossils. Yet the 1690s in particular saw the second heyday of fossil discussions, as the earlier theorising of Hooke and his opponents gave way to the pragmatic but enlightening work of classification. The culmination of this work was Edward Lhuyd's *Lithophilacii Britannica ichnographia* (1699), a classification of almost two thousand fossil types, divided into classes and accompanied by plates. Some of Lhuyd's original fossils samples survive in Oxford today in the Museum of Natural History. This Oxford edition, of which only 120 copies were printed, was heavily funded by subscriptions from the London virtuosi, including Martin Lister, Hans Sloane and Isaac Newton. Lhuyd had long been corresponding on the subject with John Woodward of Gresham College, and it is notable that Woodward's own notorious 'fix' to Burnet, his *Essay towards a Natural History of the Earth* (1695), was framed as an attempt to rewrite Burnet in such a fashion as to explain the nature and location of fossils. This unfortunate essay was rightly criticised by Lhuyd and his friends, and was still being lampooned decades later. Woodward

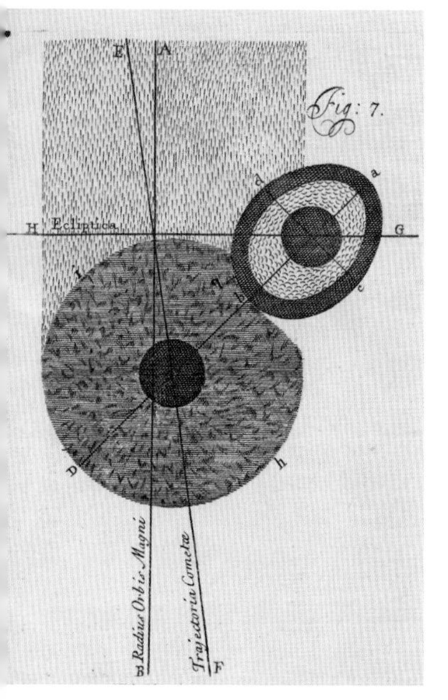


was as ever not without supporters, but once again commentators at large had more fun with than sympathy for Woodward's 'Flood', this time a providential slackening of gravity in which stone turned to soup, and fossils sank in it according to their specific gravities. Likewise with the mathematician William Whiston's Newtonian rebuttal: his *New Theory of the Earth* (1696) proposed instead that all significant catastrophes in sacred history were to be blamed this time on (again providentially guided) cometary impact.

Aubrey was a confirmed Hookian throughout all of this controversy. From the time of the *Micrographia* right up to the final disputes between Woodward and a now estranged Lhuyd, Aubrey's belief that fossils were of ultimately organic origin seems never to have swayed, and comments scattered all over his work attest to his support for Hooke's views on earthquakes and on the larger geomorphology of the earth. This is most clearly seen in his 'Digression' to *The Natural History of Wiltshire*, 'An Hypothesis of the Terraqueous Globe'. In this chapter Aubrey set down Hooke's geomorphological theory in brief with some illustrations. The current form of the world, he explained, is largely the result of earthquakes and subterranean fires, something he illustrated by redrawing some diagrams on the internal structure of the earth from Kircher's *Mundus Subterraneus*. Aubrey, like Hooke, quoted large chunks of Ovid's *Metamorphoses* to illustrate his claim, but unlike Hooke he openly stated

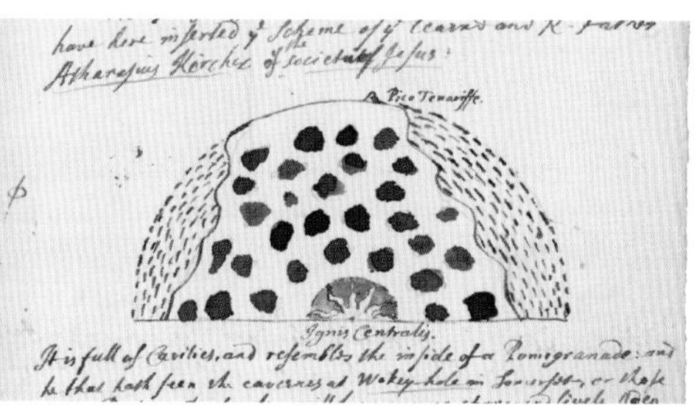


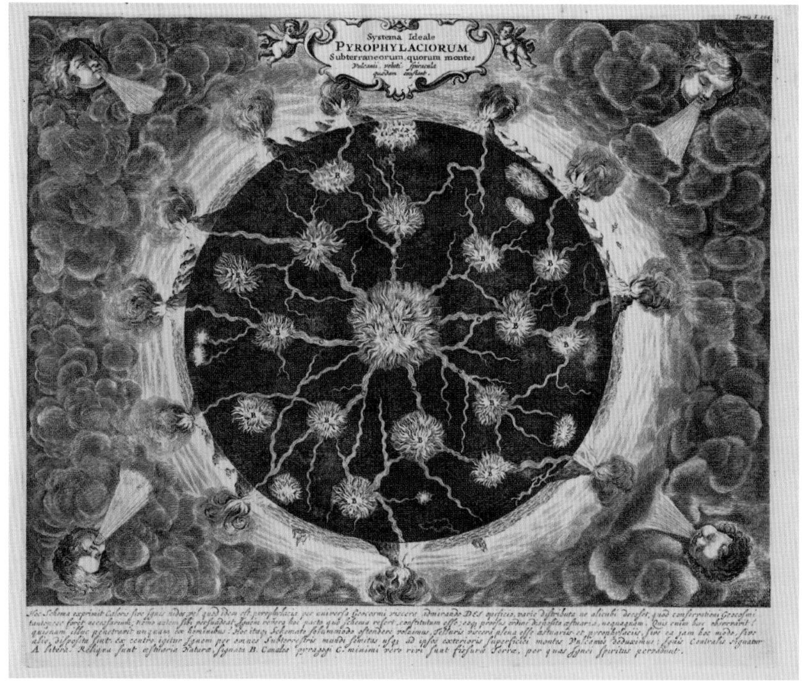


*Top* **Figure 58** William Whiston proposed that the Flood was caused by the providentially sequenced collision of the earth and a comet, illustration from his *A New Theory of the Earth from the Original to the Consummation of Things* (London 1696). Oxford, Bodleian Library, 8° Q 44 Th., fig. 7.

*Above* **Figure 59** Aubrey's 'An Hypothesis of the Terraqueous Globe' from his 'Natural History of Wiltshire' combined Kircher's ideas with those of Hooke, and repeated the claim that Steno plagiarised Hooke. Oxford, Bodleian Library, MS Aubrey 1, extract from fol. 89r.

that 'this World is much older, than is commonly supposed'. This is the corollary of Hooke's theories that Hooke himself could not voice in his public lectures. Aubrey also recorded the recent and extraordinary thesis of Edmond Halley that the earth was in fact hollow, filled not with water but with another, independently rotating planet, replete with its own population. Some sense of just how dangerous all these theories could be is suggested by Aubrey's correspondence with John Ray about the 'Wiltshire' manuscript. Aubrey hoped to publish the work in its entirety, and sent it to Ray in the form in which it now survives in the Bodleian manuscript. Ray told him to publish, but to cut the chapter on the terraqueous globe, as it was in open conflict with the book of Genesis. Aubrey angrily annotated the letter with a declaration that the whole chapter was due to Hooke's inspiration, and was the best thing in his book. Alas, Aubrey once again did not publish the manuscript, in censored or uncensored form. Hooke's own lectures were only published in 1705, two years after his death, and at a point when Hooke's reputation was entering its long, regrettable decline. Aubrey's 'Wiltshire' project, in this sense, was the one opportunity Hooke really had to get his theories on the earth's origins out into the public under the controlled pen of a disciple. It is often said that if Aubrey had published his 'Monumenta Britannica' the history of archaeology would look slightly different; this is surely also true for his *Natural History of Wiltshire* and the history of the earth sciences.

*Bottom left* **Figure 60** A very expensive book, from which Aubrey copied illustrations: Athanasius Kircher, *Mundus Subterraneus* (Amsterdam 1665). Oxford, Bodleian Library, Douce K 149, plate between pp. 194 & 195.

*Above* **Figure 61** Portrait of Edmond Halley, holding up a diagram of his system of earths within the Earth. London, Royal Society, IM/001862.

# 8

# Historical Methods

Aubrey's interests in megaliths, I have argued, were affected by larger developments in the earth sciences, developments that were opening up received chronology to new scrutiny. Aubrey's own work on megaliths, in turn, was to suggest to him new ways of doing history in more recent periods. History, as the early-modern phrase had it, has two eyes: geography and chronology, the 'where' and 'when' which underpin all historical statements. Geography itself in the period was made up of descriptive and mathematical components: the description of a given continent, country, county or place; and the mathematical treatment of such zones, from the spherical earth to the mapping of towns, forts or even single monuments.

Aubrey was interested in all of these subdivisions of history, and he probably received some instruction in all of these areas too while an undergraduate at Trinity College, Oxford. His later antiquarian work is geographical in a very precise sense: the description of a given area is called chorography, and Aubrey's work on Wiltshire and Surrey could look back to a distinguished pedigree rooted in the great Elizabethan chorographer William Camden (1551–1623), although Aubrey's chorography embraced much more natural philosophy than did the chorography of his predecessors. But it is his brief experimental ideas on chronology which will interest us here.

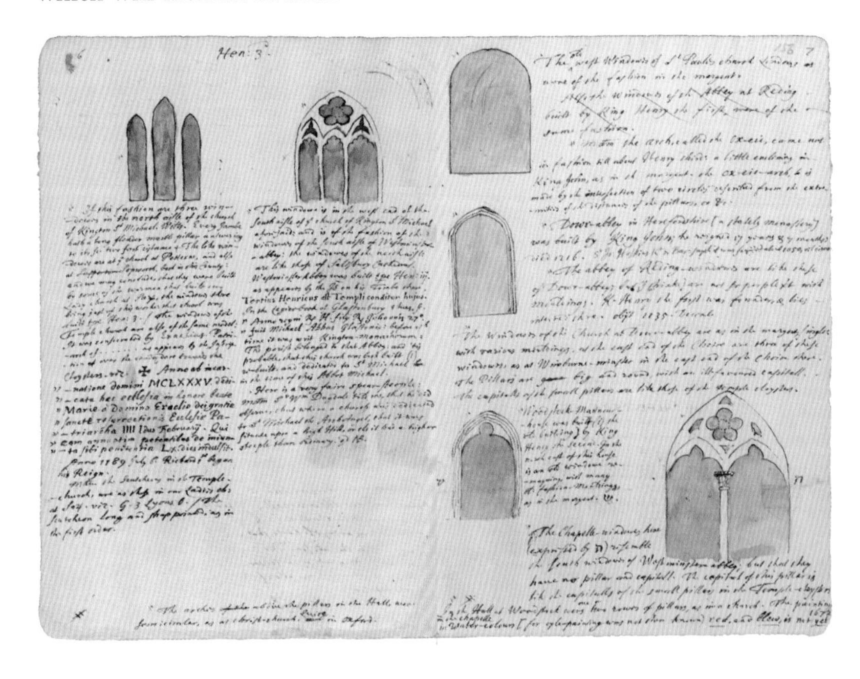

**Figure 62** Systematic dating by architectural style: Aubrey's 'Chronologia Architectonica'. Oxford, Bodleian Library, MS top. gen. c 25, fols 155v, 156r.

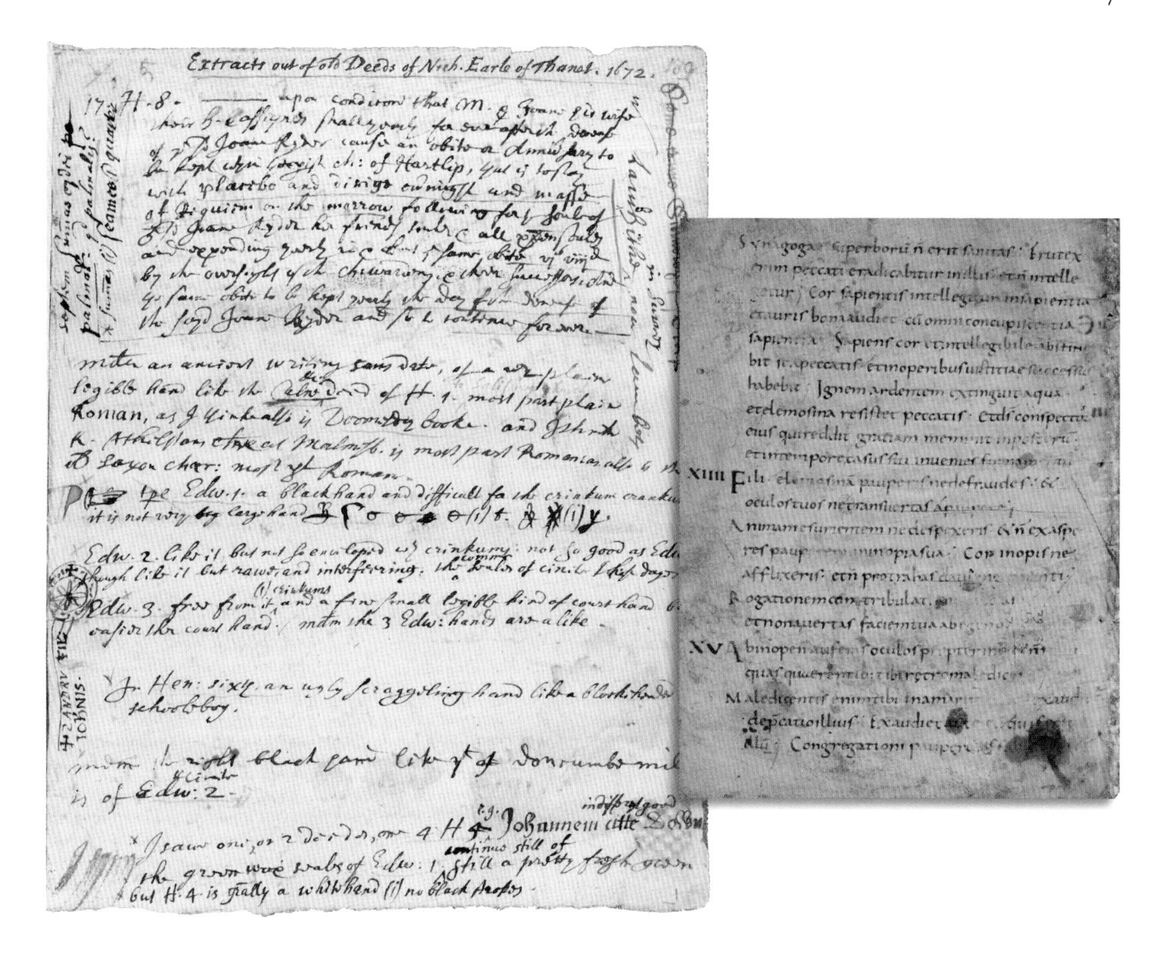
Extracts out of old Deeds of Nich. Earle of Thanet. 1672.

Towards the end of his 'Monumenta Britannica', Aubrey included a number of connected essays which he also considered to form a self-standing unit on their own. These he titled 'Chronologia Architectonica', 'Chronologia Graphica', 'Chronologia Vestiaria' and 'Chronologia Aspidologica'. Each consisted of a short, heavily illustrated essay on the chronological development of style, first architectural, then of writing, and finally of clothing and shields, the last two categories chiefly concerning themselves with the forms of illustration found in church windows and on church monuments. In short, Aubrey was proposing that monuments and even scripts can be dated if the observer has an appropriate collection of dated examples to hand, and Aubrey was providing his reader with models of such collections. The 'Chronologia Architectonica', for example, compiled in around 1671, consists of many watercolour drawings of types of arch, annotated with the name of the monarch reigning at the time of construction. Not, of course, that people had known nothing of such matters before Aubrey's time; but his significance is that he felt such knowledge required systematisation – and manuscript was perhaps the ideal form for this kind of exercise,

**Figure 63** Systematic dating by handwriting, with some facsimile attempts, and an inserted folio cut from an eleventh-century MS: Aubrey's 'Chronologia Graphica'. Oxford, Bodleian Library, MS top. gen. c 25, fols 188v, 189r.

Figure 64 Systematic dating by clothing styles: Aubrey's 'Chronologia Vestiaria', with a portrait of Chaucer. Oxford, Bodleian Library, MS top. gen. c 25, fol. 202r.

because it allowed Aubrey to write and draw complex scripts and objects on the same pages.

The 'Chronologia Graphica' is perhaps the most interesting of these attempts, if not quite as visually striking as the other essays. This is not least on account of Aubrey's own comment on the gestation of the scheme: 'I was led on to y^e^ thought, by my Chronologia Architectonica.' Aubrey used the collection of deeds owned by his friend – and, at this point, major source of charity – the Earl of Thanet, who had set Aubrey the summer job of sorting out his family's archive. Aubrey employed the content of Thanet's deeds to date the hand, and then set down several specimens of similarly dated hands, arranged in chronological order. Aubrey had long been interested in hands and even in typography, and in his antiquarian manuscripts he quite frequently imitates 'black letter' or gothic script and on one occasion Wynkyn de Worde's typeface. In the 'Chronologia Graphica' he describes hands in rather quirky fashion which would not perhaps win him high marks in a modern palaeography examination: a script from the time of Edward I, for instance, is 'a black hand and difficult for the crinkum crankum'; and from the time of Henry VI, 'an ugly scraggling hand like a blockheaded schoolboy'. ('Crinkum crankum' appears to apply to frequent loops in a script.) Aubrey even pastes into his own manuscript a dismembered leaf of a medieval manuscript, an eleventh-century page of Ecclesiasticus in the Vulgate Latin text. Aubrey's interest in the history of scripts is in keeping, too, with his larger linguistic interests, and we may also note in the 'Monumenta Britannica' manuscript an interesting few pages in which Aubrey worked through an English etymological dictionary separating and counting up words based on their language of etymological origin in order to come to some sort of statistical conclusion about the relative contributions of English's parent languages to its make-up in the late seventeenth century. His sense of the historical possibilities of language is further witnessed in his manuscripts on onomastics (i.e. the study of proper names) and English place names.

Once again, Aubrey was hardly the first English scholar to ponder the dating of hands. In 1684, for instance, on the very same day that Aubrey presented the Royal Society with the mathematical papers of Thomas Merry, the linguist and FRS Thomas Gale showed the Society a manuscript about how to make fireworks, and dated it to the time of Henry III (1207–1272) on the grounds of its content, and (explicitly) the form of its handwriting (Birch 1756–57: 4.296). Such skill had long been witnessed in English scholarship. Old English scholars in the Elizabethan period such as John Joscelin became adept at facsimile hands, and the circle of Saxon scholars gathered around Matthew Parker, Archbishop of Canterbury, also showed keen interest in the Old English hands. So the Elizabethan antiquaries were certainly expert readers of technically 'dead' scripts, and had even developed an aesthetic appreciation of the oldest scripts to the extent that a special font was developed for the printing of Old English texts, a font that very closely imitated the standard Anglo-Saxon hand, Anglo-Saxon minuscule. But the academic disciplines of 'palaeography' (the historical study of handwriting) and 'diplomatic' (the historical study of non-literary documents) did not arise until the late seventeenth and early eighteenth centuries. For a long time after such practical know-how must have been in operation, no one made a systematic effort to arrange scripts into families of related hands undergoing datable changes over time. One reason for this apparent delay may, as has been suggested, be technological – the London book trade was not as technologically sophisticated in the period as many of its continental equivalents, and an illustrated printed book would be much harder and more expensive to produce than its manuscript equivalent. Yet every new discipline needs its textbook, and the difficulty and expense of engraving plates with medieval scripts may have nipped any such venture in the bud. This hypothesis might be backed up by the existence of a manuscript tract by the Elizabethan and Stuart historian Sir Henry Spelman (1563/4–1641) named 'Archaismus Graphicus', a short Latin work on the history of character forms and on the means to determine the age of manuscripts, with a glossary of abbreviations (MS Stowe 1059). Composed in 1606 for the use of Spelman's sons, this lone manuscript is all that stands between the practical palaeography of the Parker circle in the 1560s and Aubrey's manuscripts of over a century later. It is tempting to think that Aubrey knew of Spelman's work: Spelman's showy title 'Archaismus Graphicus' ('Graphical Antiquity') is not too far from Aubrey's 'Chronologia Graphica', and we may note too that Spelman wrote a study of coats of arms, as did Aubrey, under the title of 'Aspilogia' (only published in 1654); compare Aubrey's 'Chronographia Aspidologia'. But Spelman's manuscript was not published either.

Aubrey was, however, formulating his views on the possibility of proper palaeography at an auspicious time. As a later annotation Aubrey made on his manuscript acknowledges, similar work was being undertaken at the time in France, resulting most famously in the *De re diplomatica* (1681) of the great French Benedictine monk and scholar Jean Mabillon (1632–1707). In England, throughout the 1690s the greatest of English palaeographers, Humfrey Wanley (1672–1726), was getting into his stride, and it is from this period too that the first sample books of manuscript hands on the Aubreyan model date. Wanley assisted the book historian John Bagford (1650/51–1716) in the compilation of books of specimens in dated order, and a similar manuscript assembled by the Oxonian librarian and bibliographer Thomas Hearne survives in the Bodleian (MS Rawl. Q b 4). Aubrey's skill with hands would very soon be eclipsed by the vaster talents of Wanley and Hearne, and in turn the English contribution to palaeography and diplomatic was all but buried by the huge tomes of the French Maurist monks, rooted in Mabillon's *De re diplomatica*, and continuing through Bernard de Monfaucon's *Palaeographia Graeca* (1708) and eventually C.-F. Toustain, R.P. Tassin, and J.B. Baussonnet's six-volume *Noveau Traicté de Diplomatique* (1750–65). The first English printed work that can claim to be a work specifically on diplomatic is probably the *Formulare Anglicanum* of Thomas Madox (1666–1727), published in 1702, five years after Aubrey's death. Madox's book consisted of a prefatory 'Dissertation concerning Ancient Charters and Instruments', some plates of seals and scripts, and then transcriptions of several classes of documents placed in chronological order. His major source was the archive of the court of augmentations, supplemented by documents from the Cotton Library and Corpus Christi College, Cambridge. Madox did not recognise any predecessors in the field, although he was well aware that earlier legal antiquaries such as the Puritan William Prynne had been able to date documents on palaeographical (script) rather than diplomatic (content) grounds. Mabillon's work also fed back into insular scholarship in works such as the Scottish historian James Anderson's (1662–1728) polemic and patriotic *Historical Essay Shewing that the Crown and Kingdom of Scotland is Imperial and Independent* (1705), and British scholars of the eighteenth century thereafter looked chiefly to the works of the French Benedictines and Jesuits as their foundation texts in palaeography. The brief foray of Aubrey, once again, lay neglected apart from a short but important appearance in some addenda to a work published only in 1763.

# 9

# Aubrey and Biography

Aubrey is best known today for his biographical sketches of contemporaries, the 'Brief Lives', although Aubrey did not use that title in his correspondence, preferring instead to talk merely of his 'lives' or his 'minutes of lives'. It is him we have to thank for the anecdote that the physician William Harvey 'was wont to say that man was but a great mischievous baboon'; or that the philosopher Thomas Hobbes's 'greatest trouble was to keep off the flies from pitching on his baldness'; or that the experimentalist Robert Hooke 'invented thirty different ways of flying'. Nevertheless, Aubrey's literary reputation as a biographer has proved a mixed blessing. Many have misunderstood the nature and intention of his biographical collections, diminishing his importance with the backhanded compliment that he was the greatest gossip of his age. A gossip, however, is someone who is too free with personal information, and yet Aubrey restricted his work to manuscript, and to material conveyed through private correspondence. It was the private tendency of his work, then, that allowed him to include such colour and intimacy, but his intention was not at all to cause scandal. Yet his work, even in his own age, was not entirely private, nor was it disconnected from his other intellectual interests. It is these connections we explore here.

English literary biography in Aubrey's age was neither well established nor generically stable. New ground had been broken by Francis Bacon's *The Historie of the Raigne of King Henry the Seventh* (1622). Bacon located the life of the King within the larger political arena, and eschewed providentialism, preferring instead to investigate the psychology of Henry's avarice and its effect on his public policy without any recourse to a divine plan. Bacon also appealed to manuscript sources, not just to discover material unavailable in printed sources but to get closer to Henry himself. For instance, Bacon illustrated Henry's neurotic attention to financial detail by appealing to an account book he had seen of one of the King's debt collectors, in which almost every page had been countersigned by the obsessive Henry. These details, Bacon commented, were like 'little sands and grains of gold and silver', which help 'not a little to make up the great heap and bank' of the whole work (Bacon 1998: 178). Aubrey too would collect the grains of gold and silver, and he would become an ever keener scrutiniser of personal manuscript habits. But English biography between Bacon and Aubrey for the most part remained the study of worthies. John Fuller's *The History of the*

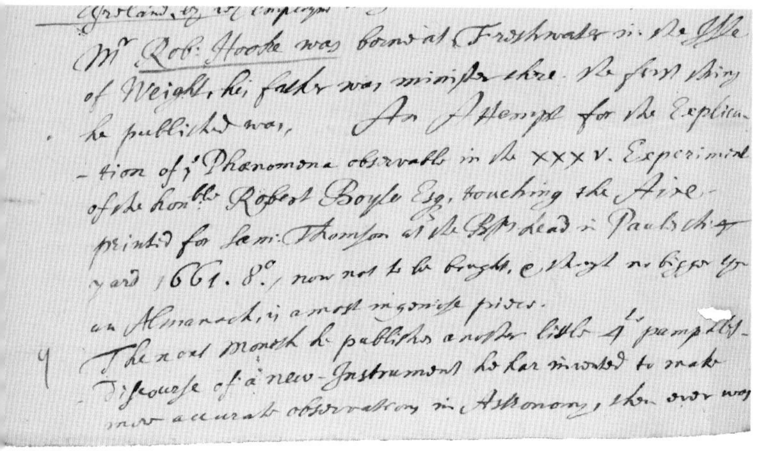
Mr Rob: Hooke was borne at Freshwater in the Isle
of Weight, his father was minister there. the first thing
he published was, An Attempt for the Explica-
-tion of ye Phænomena observable in the XXXV. Experiment
of the honble Robert Boyle Esq, touching the Aire
printed for Sam: Thomson at the Bp's head in Paul's church-
yard 1661. 8o, now not to be bought, though no bigger than
an Almanack, is a most ingeniose piece.
The next month he publish'd another little 4o pamphlet,
Discourse of a new-Instrument he has invented to make
more accurate observations in Astronomy, than ever was

**Figure 65** Aubrey writes to Wood on Hooke, one of a list of people on whom Aubrey reported; he commends Hooke's rare first pamphlet, which, 'though no bigger than an Almanack, is a most ingeniose piece'. Oxford, Bodleian Library, MS Aubrey F 39, extract from fol. 270v.

*Worthies of England* (1662) worked through the worthies county by county; although it was a useful reference work, it was neither historically investigative nor biographically colourful. Best remembered today are the lives of John Donne, Sir Henry Wotton, Richard Hooker, George Herbert and Robert Sanderson published by the angler Isaac Walton from the 1640s to the 1680s. These indeed contained many seemingly intimate details – John Donne in his final illness dressing up in his own winding sheet for his death portrait – but Walton's aim was pious, as he attempted to paint a rosy picture of English churchmanship before the decades of the civil war and interregnum.

Aubrey, to whom Walton later sent biographical details on Ben Jonson, also shared a strong but conflicted nostalgia for times past, although his frank medievalism – he occasionally dreamed of a kind of lay monasticism – had more in common with the antiquaries Sir Henry Spelman and Sir William Dugdale. But his biographical intentions differed markedly. These were developed over time from several different sources, and it is their combination which lends Aubrey's work its distinct character. First, in his biographical work as elsewhere Aubrey as ever looked back to Francis Bacon, but perhaps rather to his definition of the three parts of history in *The Advancement of Learning* (1605) than to the lengthy, polished *Henry VII*. Likening history to painting, Bacon had said in the *Advancement* that history, like a picture or an image, could be unfinished, perfect or defaced. An unfinished history is a rough draft; whereas 'ANTIQVITIES are Historie defaced, or some remnants of History, which haue casually escaped the shipwrack of time' (Bacon 1951: 86 (2.II.1)). Unfinished history is therefore simply the product of an unfinished historian; whereas the antiquarian deals with inherently imperfect sources. Bacon saw nothing deficient with this kind of history; 'any deficience … is but [its] nature'. The diligent follower of antiquities must therefore collect widely, from many different kinds of source: 'out of monuments, names, words, proverbs, traditions, private records and evidences, fragments of stories [i.e. histories], passages of books that concern not story, and the like'.

Aubrey would do all of these things, but in his strictly biographical work he also functions as a source of 'Antiquities' himself, creating further 'remnants' by recording things that he knew was information ac-

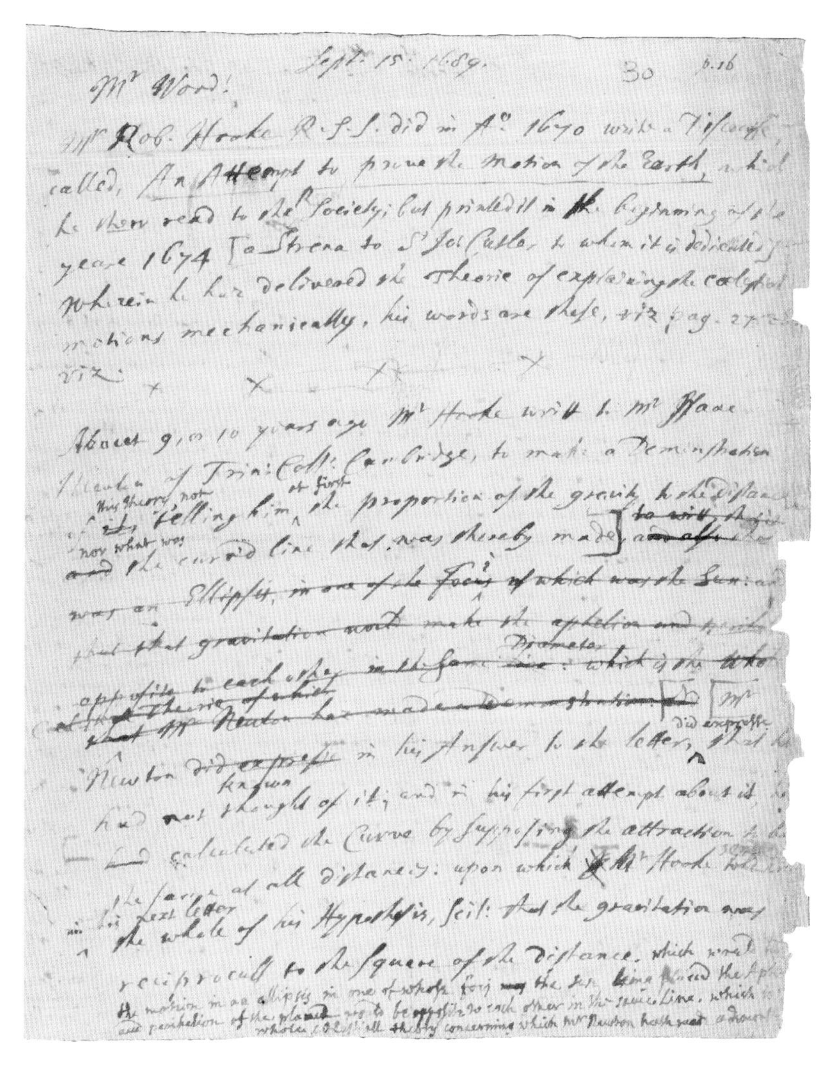
Sept. 15: 1689.

Mr Wood!

Mr Rob. Hooke R.S.S. did in Ao 1670 write a Discourse called, An Attempt to prove the Motion of the Earth, which he then read to the R. Society; but printed it in the beginning of the yeare 1674 [to Strena to Sr Jo. Cutler to whom it is dedicated] wherein he hath delivered the Theorie of explaining the coelestiall motions mechanically, his words are these, viz pag. 27 viz:

About 9 or 10 yeares ago Mr Hooke writ to Mr Isaac Newton of Trin. Coll. Cambridge, to make a Demonstration of this theory, not telling him, at first the proportion of the gravity to the distance nor what was the curved line that was thereby made

Newton did in his Answer to the letter, did expresse that he had not thought of it; and in his first attempt about it, calculated the Curve by supposing the attraction to be the same at all distances: upon which Mr Hooke sent in his next letter the whole of his Hypothesis, scil: that the gravitation was reciprocall to the square of the distance.

Figure 66 Aubrey writes again to Wood on Hooke; this draft, in which Aubrey claims that Hooke rather than Newton came up with the inverse square law, has been emended by Hooke, whose overseeing hand is clearly visible. Oxford, Bodleian Library, MS Aubrey 6, fol. 30r, 15 September 1689.

quired by or known uniquely to him. Aubrey also leaves us in no doubt that it was to Bacon's discussion of antiquities that he looked, because he used the 'planks from a shipwreck' metaphor several times throughout his own writings. As he stated in the preface to his life of Hobbes, the lengthiest and arguably the most polished of all his biographies, 'I hu*m*bly offer to y$^{e}$ *pre*sent Age & Posterity, tanquam Tabula*m* Naufragij ['as a plank from a shipwreck']; & as plankes & lighter things swimme, & are *pre*servd, where the more weighty sinke & are lost' (MS Aubrey 9, fol. 29r). Aubrey here was being doubly Baconian, for in an elegant modesty topos he alludes also to Bacon's essay 'Of Praise': 'Certainly, fame is like a river, that beareth up light things and swollen, and drowns things weighty and solid'. Aubrey clearly associated the 'tanquam tabula' tag with his biographical work in particular, for he later inscribed it as the epigraph to his 'Lives' manuscripts. Nevertheless, Aubrey did not doubt the lasting value of his biographical collections, as we shall see.

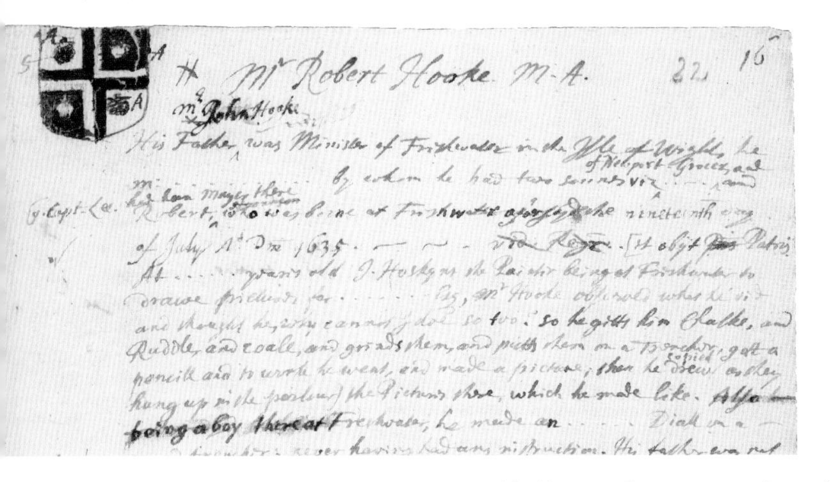

**Figure 67** The opening of Aubrey's own life of Hooke, with his habitual lacunae and emendations. Oxford, Bodleian Library, MS Aubrey 6, fol. 32r.

A further desire that shaped the making of Aubrey's biographical work was the project to reform astrology. Bacon again had called for the purification rather than the outright rejection of astrology, although he made his hostility to horoscopes plain. Aubrey, however, had high hopes for astrology, as reflected in his manuscript 'Collectio Geniturarum', or 'collection of genitures', a set of horoscopes of eminent men. As he wrote to Wood in 1671, 'I have and shall get 30 Nativities exactly, many of them of men famous in Learning, Wealth and valour, etc.' (MS Wood F 39, fol. 131r). Aubrey eventually produced a vellum-clad notebook dated 1677 of over sixty such nativities, although no longer restricted solely to people of eminence (MS Aubrey 23). This astrological dimension was also incorporated into his biographical manuscripts, although it is all but obscured by a tradition of modern editorial hostility to any astrological content. But the manuscripts themselves tell a quite different story, with many of the lives surmounted by or surrounding a horoscope evidently drawn in at the beginning and not the end of composition. Sometimes Aubrey even drew a blank horoscope, hoping to obtain the relevant details on time of birth at a later date. Aubrey's hope was evidently that a large collection of biographies with astrologically accurate horoscopes could be used to test the congruence between astrological prediction and the actual course of the subject's life. This is probably the area of Aubrey's biographical work that has fared least well, and indeed has been obscured by editorial excision. But astrology was still accepted as a component of medical diagnosis by many at the time, and Aubrey was particularly interested in the papers of the eminent astrological medic and preacher Richard Napier (1559–1634), which had by that time been acquired by Aubrey's friend Elias Ashmole, himself of course a keen astrologer. Other astrologer friends included John Gadbury (1627–1704) and Henry Coley (1633–1704), and the latter, who cast Aubrey's horoscope in 1671, was able to tell Aubrey that most of his misfortunes could be blamed on the stars, something Aubrey was perhaps perversely relieved to hear. Coley also supplied Aubrey with a horoscope for Hobbes, which Aubrey pasted into the inner cover of his life of Hobbes. If astrology now seems the least successful of Aubrey's biographical researches, it is worth pondering whether the bulk of Aubrey's papers would ever have ended up in the Ashmolean, preserved for modern readers, were it not for the shared interests of Aubrey and Ashmole himself.

Yet in terms of the gestation of the lives themselves, the decisive influence here was the Oxford historian and antiquary Anthony Wood (1632–1695). Of all the scholars associated with Aubrey, it is Wood who exercised the greatest influence, eliciting a stream of biographical information from Aubrey, whom Wood repaid by failing to acknowledge him in his printed work, mutilating his manuscripts, publishing material Aubrey had declared *inter nos*, and abusing his good name. They met in 1665 or 1667, in Oxford. Aubrey had been at Trinity College with Wood's elder brother, and had recently encountered a new history of the University and its worthies by the Corpus Christi College antiquary William Fulman, the *Academiæ Oxoniensis notitia* (1665). Aubrey was initially informed that the work was by Wood, and so Aubrey presented him with a copy marked up with Aubrey's additions (now Wood 614(1)). Thus began a close friendship that lasted a quarter of a century, before going disastrously wrong. Nevertheless, it is probably thanks to Wood that we have Aubrey's biographies at all, for Wood, who began working on his classic *Athenæ Oxonienses* a few years after their initial meeting, soon began to use Aubrey as his main biographical consultant and fact-finder. *The Athenæ Oxonienses* was to be a biographical encyclopedia of all the writers and bishops educated at Oxford between 1500 and 1690, no matter in what field they worked, each entry supplied with a bibliography of published and unpublished works. The *Athenæ Oxonienses* finally appeared in two large folio volumes in 1691–92, was frequently added to first by Wood himself, and then by a chain of antiquaries right down to the early nineteenth century. With the appearance of Philip Bliss's thereafter standard edition of 1817, the work had swelled to four volumes, and was perhaps the major single source underpinning the earlier entries in the old *Dictionary of National Biography*, and more recently its full revision, *The Oxford Dictionary of National Biography*. Wood and, through him, Aubrey are therefore the grandfathers of the idea of a national biography.

Nevertheless, Aubrey's own lives are independent from Wood's final entries to the *Athenæ*, and are much more vivid than Wood's comparatively dry distillations. From the early stages of their collaboration, Aubrey researched and answered strings of enquiries from Wood, often by searching out surviving relatives or colleagues of Wood's subjects, and returning answers to Wood, who would then docket the letters and archive them. These letters, still surviving in a fairly complete run in Wood's papers, show how Aubrey was unable and unwilling to keep business and pleasure separate. Aubrey listed facts on the lives, careers and publications of others, while simultaneously keeping up his

**Figure 68** The philosopher Hobbes presented his own books to 'my noble frend M[r] Aubrie'; this inscription is from Hobbes's *Examinatio & Emendatio Mathematicæ Hodiernæ* (London 1660). Oxford, Worcester College, donation inscription.

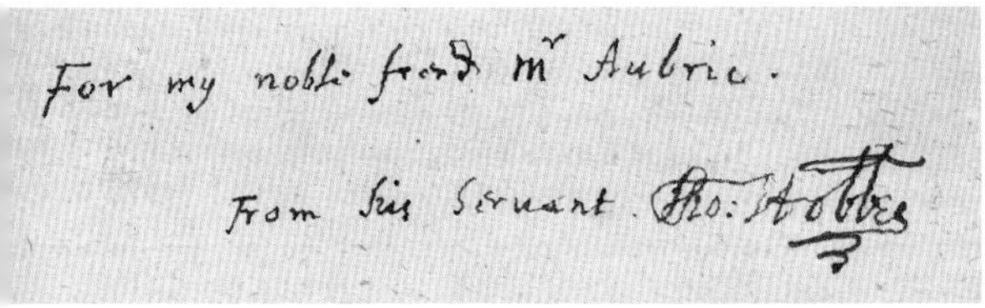

own personal friendship with Wood, whom he frequently addressed as 'Anthony', 'O Anthony!', 'Dear Anthony', an interesting intimacy at a time when even close friends usually used surnames and titles in their familiar correspondence and even in their diaries. It was in the course of this correspondence that Aubrey had the idea – something that Wood encouraged – of writing his own short biographies rather than just collecting materials for Wood. The idea arose out of Aubrey's notes on Hobbes, the philosopher he most admired, and whose friendship he cherished. Aubrey sent extensive notes on Hobbes to Wood, as Hobbes had attended Magdalen Hall *c.* 1603–08. Hobbes was later to attack his Oxford education bitterly, and Wood did not approve of Hobbes as a political philosopher, as his eventual seven-column entry on Hobbes makes clear. But Aubrey had a great capacity for hero-worship, and Wood did not discourage him from venturing to organise his notes and publish his life of Hobbes as a separate work. Aubrey enlisted the aid of a young academic to translate the work into Latin, which both Aubrey and Wood clearly regarded as something in which they both had a stake. The young translator, Richard Blackburne, frequented the London coffee shops with the virtuosi and was well known to Aubrey's specific circle of Royal Society friends. But having given Blackburne his text, Aubrey then suddenly decided to write another life, this time to remain in English, and not to repeat anything in the Blackburne text. Aubrey reported this to Wood without any sense that Blackburne's work was being competed with or criticised. Rather, Aubrey simply marvelled at his own productivity, and predicted a completed life of eight sheets, probably eight quarto gatherings, which would total 64 pages. It was never actually printed, but a manuscript copy survives in Aubrey's papers (MS Ballard 14, fol. 124r; MS Aubrey 9). Aubrey then resolved to write more such lives, specifically of three other of his virtuosi friends whom he adulated: Sir William Petty, Sir Christopher Wren and Robert Hooke. Aubrey cannot have regarded these new efforts as entirely private, as he mentions that he will ask Petty to review his own life – Petty subsequently did so, 'and would have all stand' – but Aubrey had eschewed the idea of printing such a life or even keeping it himself, merely sending it to Wood where 'it shall be left with your papers for Posterity hereafter to read'. Aubrey then took a further step, wrote out a 'Kalendar' of fifty-five people worthy to have their lives written, and set out to do so. Over the coming months he swelled his manuscript collection of 'Lives', widening it to include people who had been long since dead and cannot ever have been known to Aubrey. All of these initial stages happened between early 1680 and early 1681.

Aubrey's collection was increasingly eclectic, but it is nevertheless notable how many of his lives are of the virtuosi – Hobbes, Petty, Wren, Hooke, but then subsequently dozens more, especially of the English mathematicians, a subcategory Aubrey also later thought about pursuing as a stand-alone collection, and on whom he had long been collecting notes. One of his intentions here was to honour the English mathematical tradition to which he felt he too belonged. But another intention is equally significant. Aubrey's desire to write the lives of the most prominent of the experimentalists of the time – and often to show them to their subject (e.g. Petty, Hooke, Pell) – arose out of his realisation that he could employ his biographical collections to witness to the right side in the increasingly common disputes in the nascent scientific community about what we now would term priority and intellectual property. Aubrey himself would grow increasingly concerned to protect his own historical, chorographical and archaeological work. Particularly in the cases of friends like Francis Potter the Trinity College experimentalist or Robert Hooke in his acrimonious disputes with Isaac Newton, Aubrey's notes defend the interests of his friends. Nowhere is this more obvious than in his letters to Wood about Hooke. Aubrey clamoured (in vain) for Wood to do right by Hooke, and indeed the draft letter from Aubrey to Wood claiming priority in the theory of universal gravitation most shrilly is extensively annotated and in places wholly scripted in the hand of Hooke himself. Aubrey had discovered a new, polemical use for biography, and this aspect was directly powered by the new kinds of problem facing the experimentalists about who owned scientific discoveries, and at what point in their development.

Aubrey's own attitude to the style appropriate to biography also explicitly emerged in the years covering the life of Hobbes and then the subsequent 'Booke of Lives'. Coming to resent Blackburne's usurpation of the life of Hobbes, he came to resent Blackburne's style itself, and in so doing to formulate his own:

> Mr Dryden and My Lord Jo Vaughan much approve of it; but for the compiling, they two agree to leave out all minute things; there will be the trueth, but not the whole. They will not mention his [i.e. Hobbes] being a Page. Sir W. Petty perused my Copie all over and would have (as A[nthony] W[ood]) all stand. But I must submit to these great Wits; but in the meantime I suffer the grasse to be cutt under my feet; for Dr Blackburne will have all the Glory. 'Tis writ it in a high style. Now I say the Offices of a Panegyrist and a Historian are much different. A Life is a short History and there the Minuteness of a famous person is gratefull. (MS Ballard 14, fol. 131, 27 March 1680)

Aubrey also resolved not to repeat material already published on his subjects. To some extent, his attitudes to language here can be paralleled with the often-repeated claims from Royal Society apologists that they would employ a purified language in all their work, 'preferring the language of Artizans, Countrymen, and Merchants, before that, of Wits, and Scholars' (Sprat 1667: 113). Aubrey's manner of composition, too, was bound to lead to a choppy text as long as it remained in manuscript. He tended to write at first the details he had to hand, but left gaps for names, dates or numbers to be filled in at a later date, and his frequent additions, interlineations, cancellations and lacunae render his lives peculiarly difficult to turn into continuous modern printed text. Nevertheless, although Aubrey may have developed habits best suited to manuscript composition, he did not mean his efforts to lie hidden for ever. His self-deprecatory remarks envisage a future audience; and his letters to Wood are full of reflections on the value his lives will possess '50 yeares hence', and how he looked to Wood to preserve his work until it was safe to be published. Wood did not do so; the men quarrelled over accusations of *scandalum magnatum* (the defamation of peers) levelled against Wood; and Wood cut away from the lives' manuscripts a sheaf of Aubrey's lives, now forever lost, fearing that the papers would be seized and held against him. As we have seen, Aubrey eventually managed to get his manuscripts deposited in the Ashmolean, now safely out of his one-time collaborator's grasp.

Aubrey's lives, we noted, have long been described as gossipy, a projection of the modern literary taste for the quirky back onto materials that did not quite intend that kind of effect. Nevertheless, it has to be granted that Aubrey's sense of the domestic detail, his informal vocabulary, his fondness for putting direct speech into the mouths of his subjects, and not least his sense of humour do indeed make his lives overflow in ways that cannot be suppressed by fixing their origins variously in the Baconian ideal of the antiquary, the search for a comparative astrology, the desire to defend the achievements of one's more significant friends, and a sense that the writer of a 'short History' should avoid panegyric and embrace detail. Aubrey's lives have had a vigorous afterlife following their first printed publication because Aubrey proved to be an early writer of a style of biography that came to displace the earlier mode of the exemplary life. We prefer Aubrey's versions of his contemporaries, and that is why Aubrey is a literary phenomenon today in a way that Wood is not. Nevertheless, given the larger intentions of this study, I have sought to explain the contemporary contexts and origins of Aubrey's biographical work in order to make him intelligible in terms his contemporaries would understand. Not only does this put Aubrey's lives back in dialogue with his other scholarly interests, but it may also suggest new ways of thinking about the origins of modern biography itself.

# Conclusion

This book has examined the role of one man in several related seventeenth-century discussions. Aubrey's involvement with scientific clubs, their repositories, and their interests in mathematics, the possibility of a reformed education and an artificial language, the emergent earth sciences, the study of ancient monuments, and even of scripts and church windows, have been surveyed. Aubrey's own distinctive contribution to the study of megaliths and to the writing of the histories of places and people has also been stressed, as these are his lasting personal achievements. But I have treated Aubrey as a scholar among scholars, and it is the group phenomena in which Aubrey participated which are the real subject here. I have spent more time than is customary emphasising the institutional innovations Aubrey experienced and assisted, because these seem to me at least as important as the 'pure' ideas themselves, wherever they are deemed to reside. Science, indeed all intellectual pursuit, is not just a set of discoveries or hypotheses, but the institutionalising of the means of generating, testing, discussing and storing these ideas. Aubrey was also working right on the cusp of the next necessary step in this process – professionalization. Aubrey may have been a pioneer in many areas, but he was also among the last of a type of interested amateur that had begun the slow process of dying out.

At the outset it might have seemed odd to expect all Aubrey's disparate interests to be associated. I hope that I have showed how to Aubrey's mind, and to the minds of his contemporaries, they indeed were. What we now call 'science' has since shed many of these interests, and for that reason little that Aubrey achieved looks 'scientific' today. He made no serious contribution to an activity he loved: mathematics. He laboured to force his friends to perfect something linguists today find redundant: an artificial language. He tried to date texts on the basis of handwriting, now clearly the job of a historian not a 'natural philosopher': palaeography. But unless we have a model of the origins of modern science suitably elastic to see why these activities were relevant to the larger movement of scientific reform, we will not understand how or indeed why modern science came into being. Aubrey assists us here, because he was committed to areas both inside and outside the currently agreed domain of science.

But is there any principle that underlies all Aubrey's interests, regardless of whether we choose to see them as 'scientific' or not? The first caution we must heed is that there is no reason why a person's interests must be controlled by a master idea, one commitment that underlies all activity. Nor is any individual truly representative of their age. Nevertheless, we can still identify some broad principles that held together Aubrey's thought, and that do tell us something about intellectual endeavour in his day. First, he was interested in comprehensive collection, a Baconian passion, but one toughened up by the mathematicism Aubrey had imbibed from Cartesian and Hobbesian philosophy. Even Aubrey's interest in horoscopes, something few of his colleagues in the Royal Society shared, can be seen to flow from an interest in method: Aubrey was beguiled by the determinism that location and birth-time might enforce, and to that end he was willing to correlate behavioural traits with local soil conditions, and to collect a large number of horoscopes so that a comparative study might be undertaken. Opinions differed on astrology sharply at the time. Hooke, fond though he was of Aubrey, would have considered this latter activity to be both eccentric and irrational. Ashmole would have considered it to be reasonable and laudable. There was no one received view of astrology, either in favour or against, although Hooke's view has of course come to prevail.

The second broad principle powering Aubrey's interests is the idea of method as applied to historical research. This drove Aubrey to bring practical mathematics to Stonehenge, and to accept an age for such monuments that was hard for most of his contemporaries to stomach, obsessed as they were with doing history from texts, with objects holding a secondary status. I think this mindset is not only directly connected to the Royal Society's public mantra of putting objects before texts, but also a counterpart to the geological theories being developed by Hooke and Halley, and it is important that Aubrey accepted the general thrust of their work. If humans left monuments before any texts existed, if fossils represented species no longer among us, if mountains suggest that the earth is much older than was commonly accepted, then for Aubrey these conclusions stood, regardless of the weight of received tradition. And all these conclusions were to do with stretching time, and with broadening the categories of evidence we admit when discussing it.

'Received tradition' here is biblical tradition, and this brings me to a slightly surprising final observation: Aubrey's almost complete lack of interest in theology. He shared this trait with Hooke. In an age in which many of the nascent scientific community were engaged with the religious implications of their work, Aubrey's silence is interesting. As we have seen, he had little patience with Ray when the latter criticised his geological theories as unbiblical. Aubrey was not a heretic, to be sure – he just didn't care very much, and heresy, as John Milton pointed out, means choice and commitment. Churches were for Aubrey troves of history, not temples for worship. Aubrey admired the monastic life, but only because he envied the peace and quiet. He did not despise the Pope, but only because he did despise the Protestant iconoclasts for wrecking his cherished monuments. Some of the anxieties of the more religiously attuned scientists about potential heterodoxy are, therefore, well founded, not simply because some experimentalists were indeed embracing heterodox religious ideas, but because others, too, were just not sufficiently interested in continuing to be accountable to what had up until that point been the undisputed intellectual policeman, theology. And a lack of interest is in the end a more effective killer of ideas than outright opposition. This potential disconnection between science and religion and the anxieties it has provoked have had an interesting subsequent history, and one that is far from over.

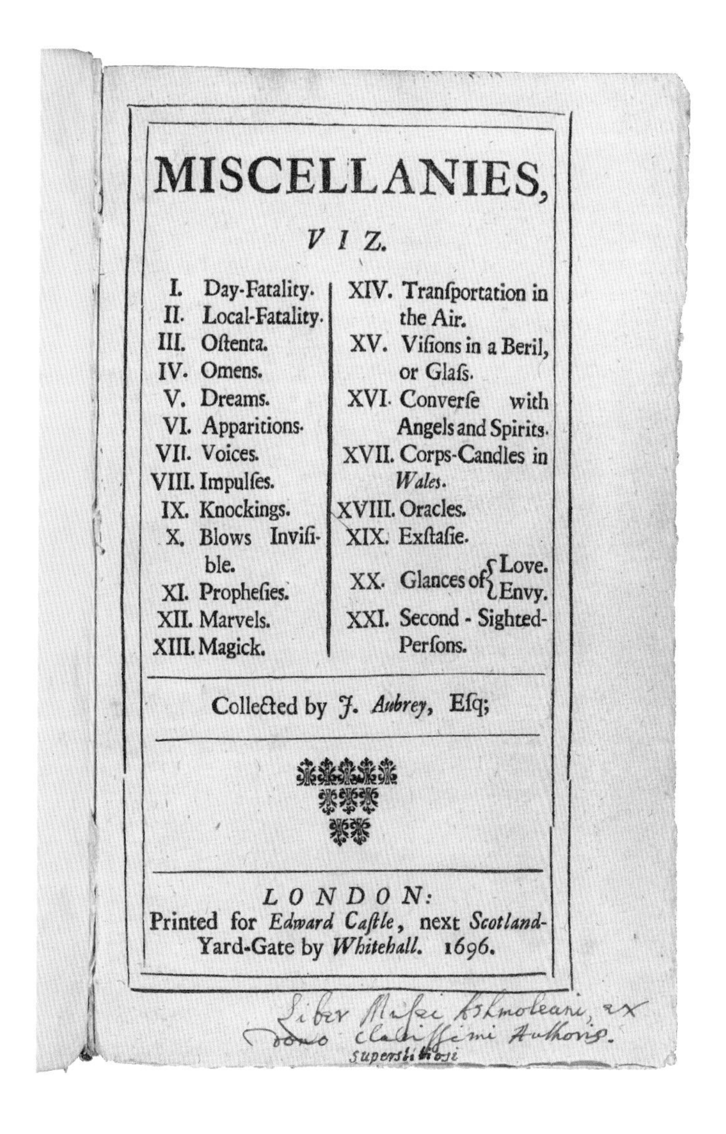
MISCELLANIES,

*VIZ.*

I. Day-Fatality.
II. Local-Fatality.
III. Oftenta.
IV. Omens.
V. Dreams.
VI. Apparitions.
VII. Voices.
VIII. Impulfes.
IX. Knockings.
X. Blows Invifible.
XI. Prophefies.
XII. Marvels.
XIII. Magick.
XIV. Tranfportation in the Air.
XV. Vifions in a Beril, or Glafs.
XVI. Converfe with Angels and Spirits.
XVII. Corps-Candles in *Wales*.
XVIII. Oracles.
XIX. Exftafie.
XX. Glances of { Love. Envy.
XXI. Second - Sighted-Perfons.

Collected by *J. Aubrey*, Efq;

*LONDON:*
Printed for *Edward Caftle*, next *Scotland-Yard-Gate* by *Whitehall*. 1696.

Liber Musæi Ashmoleani, ex dono clarissimi Authoris.
superstitiosi

**Figure 69** The only work Aubrey ever managed to publish solely under his own name, his *Miscellanies* (London 1696). Oxford, Bodleian Library, Ashmole E 11, title page.

APPENDIX

# Aubrey's Books

Figure 70 Isaac Newton's landmark *Principia Mathematica Philosophiæ Naturalis* (London 1687); this copy was gifted to Aubrey by Thomas, Earl of Pembroke in 1690, and then gifted again by Aubrey to the Ashmolean. Oxford, Bodleian Library, Ashmole F 21, title page.

Aubrey collected books throughout his life. As a young man called away from Oxford at the start of the civil war, Aubrey recalled journeying back to the country with a copy of Thomas Browne's freshly published *Religio Medici* in his pocket, a celebrated book 'which first opened my understanding'. Other student purchases came from the sale of the library of the poet William Cartwright. In later life Aubrey spent many hours in the second-hand markets and auctions of London with Hooke, who himself amassed a huge personal library in Gresham College, as well as tending to the Royal Society's books. Indeed, Gresham College in the late seventeenth century housed several large private libraries, as resident Gresham Professors such as John Woodward maintained their own collections. By the turn of the century there must have been well over 10,000 books on the site, and Gresham was where Aubrey spent much of his time.

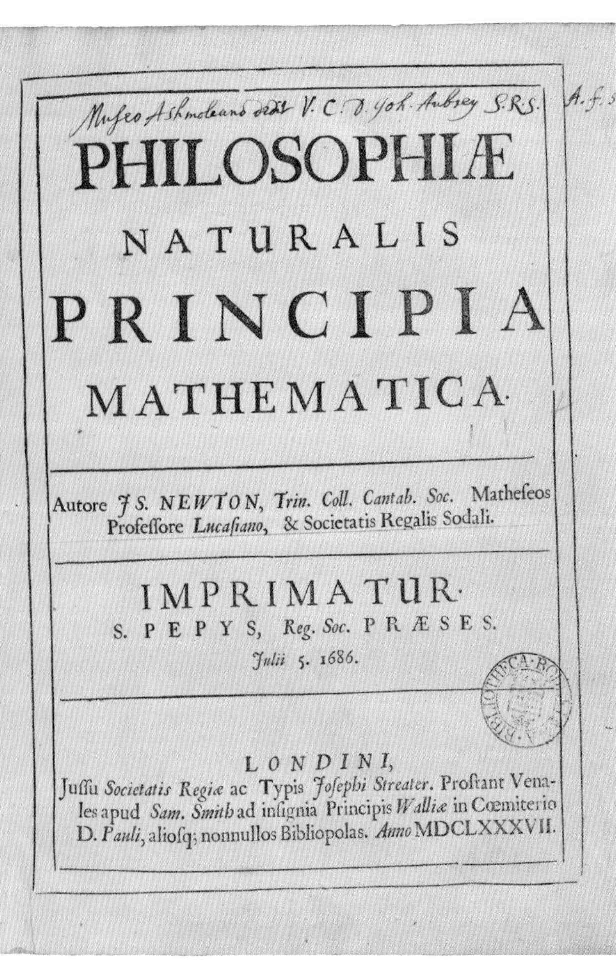

PHILOSOPHIÆ
NATURALIS
PRINCIPIA
MATHEMATICA.

Autore *J S. NEWTON*, *Trin. Coll. Cantab. Soc.* Matheseos Professore *Lucasiano*, & Societatis Regalis Sodali.

IMPRIMATUR.
S. PEPYS, *Reg. Soc.* PRÆSES.
*Julii* 5. 1686.

LONDINI,
Jussu *Societatis Regiæ* ac Typis *Josephi Streater*. Prostant Venales apud *Sam. Smith* ad insignia Principis *Walliæ* in Cœmiterio D. *Pauli*, aliosq; nonnullos Bibliopolas. *Anno* MDCLXXXVII.

Throughout his life Aubrey gave away many of his books too. In Oxford, as we have seen, he donated a small mathematical library to Gloucester Hall and a much larger set of books to the young Ashmolean Museum, and he also donated directly to the Bodleian Library and to New Inn Hall. In London, he gave books intermittently to the Royal Society Library. His gifts were perhaps not always uncontroversial: when he tried to extract his copy of Hobbes's notorious *Leviathan* from Anthony Wood in order to give it to New Inn Hall library, for instance, the latter taunted him, predicting 'they will not take [it]. Doe you think that it is a booke for young students to study in? No, rather to make them Hobbists and confound them.' But Wood and Aubrey had fallen out by this point, and Wood in his railing at Aubrey's generosity was being unfair. For Aubrey continued to acquire and to donate books even as he slid further into debt. Indeed, he was reduced to selling his books too, although to appreciative friends. In mid-1674, for instance, Hooke recorded in his journal a long string of fascinating titles that he had just acquired from Aubrey:

> Received from Mr. Aubery, Pappus, Apollonius Pergaeus, Diophantes, Copernicus, Bacon *De mirabilibus artis et naturæ*, Geber *Alchimia*, Lull *testamentum*, S. Hartlib about engines and husbandry, Neipeir *Logarithms* Brerewood *de ponderibus*, Andersons [mathematical tracts], Descartes *de Lumine*, Pell in high Dutch, Pecquett, Galileo *de Regula proportionis*. (Hooke 1935, for 6 July 1674, adjusted from edition)

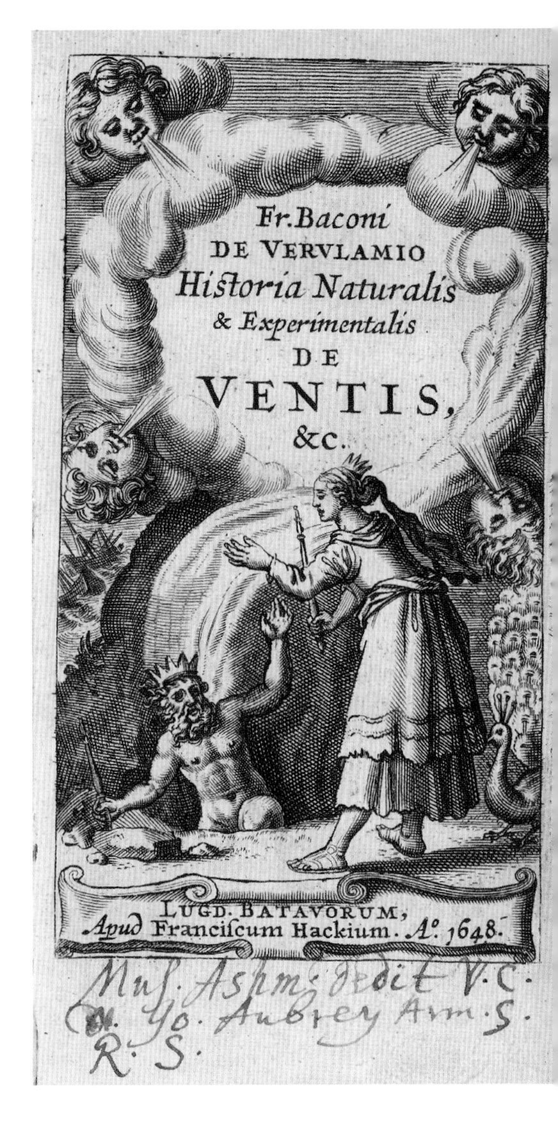


Figure 71 Aubrey's presentation copy to the Ashmolean of Francis Bacon, *De ventis* (Leiden, 1648). Oxford, Bodleian Library, Ashmole A 42, title page.

This is an extraordinary list of mathematical and scientific books, and this set of fifteen titles would today raise hundreds of thousands of pounds at auction, although Hooke, we may guess, acquired them from bankrupt Aubrey at a snip.

Aubrey never made a catalogue of his books, and indeed he probably never possessed a collection stable enough to be called a 'library'. He had a store of books in the country, and he stashed much of his current reading with urban friends, expecting to be able to call by and pick up his books when he felt the need. In his final decades he had little choice: Aubrey's delitescence precluded him from maintaining a fixed collection, as he had no house in which to put it. Nevertheless, we can still establish a surprisingly long and detailed list of books Aubrey owned, either because he discussed such books in his correspondence, or because he gave his own copies to institutions where they can still be identified today. More than 250 extant copies can be traced back to Aubrey, and there are hundreds of further titles that he must have owned at some point, even if we cannot locate his own copies today. Now and then signed copies turn up on the second-hand market. On these pages are illustrated notable examples of Aubrey's books found in Oxford libraries today, including books by Aubrey, books presented to Aubrey by their authors, a manuscript copied by Aubrey. There is also a fourteenth-century manuscript presented by Aubrey to the Bodleian Library on the frontispiece of this book.

*p.104* **Figure 72** Aubrey's presentation copy to Gloucester Hall of Thomas Digges, *A Geometrical Practical Treatize named Pantometria* (London, 1591); although heavily defaced, the original signature is that of Ben Jonson. Oxford, Worcester College, title page.

A GEOMETRICAL PRACTICAL TREATIZE NAMED PANTOMETRIA, diuided into three Bookes, LONGIMETRA, PLANIMETRA, and STEREOMETRIA, Containing rules manifolde for menſuration of all *Lines*, *Superficies* and *Solides*: with ſundrie ſtrange concluſions both by Inſtrument and without, and alſo by *Glaſſes* to ſet forth the true Deſcription or exact Platte of an whole *Region*. Firſt publiſhed by *Thomas Digges* Eſquire, and Dedicated to the Graue, Wiſe, and Honourable, Sir *Nicholas Bacon* Knight, Lord Keeper of the great Seale of England. With a Mathematicall diſcourſe of the fiue regular *Platonicall Solides*, and their *Metamorphoſis* into other fiue compound rare *Geometricall Bodyes*, conteyning an hundred newe *Theoremes* at leaſt of his owne *Inuention*, neuer mentioned before by anye other *Geometrician*.

*LATELY REVIEWED BY THE* AVTHOR himſelfe, and augmented with ſundrie *Additions*, *Diffinitions*, *Problemes* and rare *Theoremes*, to open the paſſage, and prepare away to the vnderſtanding of his Treatize of *Martiall Pyrotechnie* and great *Artillerie*, hereafter to be publiſhed.

AT LONDON
Printed by *Abell Jeffes*.
ANNO. 1591.

# Authorities and Further Reading

The standard work on Aubrey is still Hunter 1975. There are good biographies by Powell (1963) and Tylden-Wright (1991), although I have not further referenced these below. Balme 2001 presents some of the Aubrey–Wood correspondence. There is still no edition of the complete correspondence, although one is now under way. Kate Bennett's edition of the *Lives* is forthcoming with Oxford University Press, and will contain definitive commentary.

Chapter 1 See Syfret 1950; Hunter 1975: ch. 1, 1981, 1994; Feingold 1984, 1997.
Chapter 2 See Turnbull 1950, 1952; Hoppen 1970: 210–11; Frank 1973, 1980; Feingold 1997, 2005; Webster 2000: esp. 54–7, 144–78, 218; Shalev 2002.
Chapter 3 See Bennett, Johnston and Simcock 2000; Impey and MacGregor 2001: chs 18 and 19; MacGregor 2001; Webster 2000: 153–78, 320–21; Moorhead 2001.
Chapter 4 See Bennett 2001: 242–4; Feingold 1997; Frank 1973: 194, 198–201; Hunter 1975: 47–55, 244–5; Malcolm and Stedall 2005: 238–42; Powell 1963: Appendix B; Turner 1973; Stedall 2002; Bennett 2009.
Chapter 5 See Turner 1978; Cram 1994; Bennett and Mandelbrote 1998; Cram and Maat 2001; Lewis 2001, 2007; Henderson and Poole 2010.
Chapter 6 See Hutton 2009: 66–73, 125; Burl 1979: 41–6; Piggott 1985: ch. 4, 1989: ch. 4; Ucko et. al. 1991: *passim* but esp. 8–59; Parry 1995: 275–307; Seaton 1935: 236–42; Haycock 2001: ch. 5.
Chapter 7 See Hunter 1975: 84, 100, 221–2; Parry 1995: ch. 10; Hazelgrove and Murray 1979; Rudwick 1985: chs 1 and 2; Kollerstrom 1992; Poole 2006, 2008, 2010: chs 5 and 7.
Chapter 8 See Hunter 1975: 156–57; Fellows-Jensen 2000.
Chapter 9 Hunter 1974; Bennett 1993, 2007, 2009; Balme 2001.
Appendix Gunter 1930, 1931; Buchanan-Brown 1974; Kelly 1981; Powell 1950.

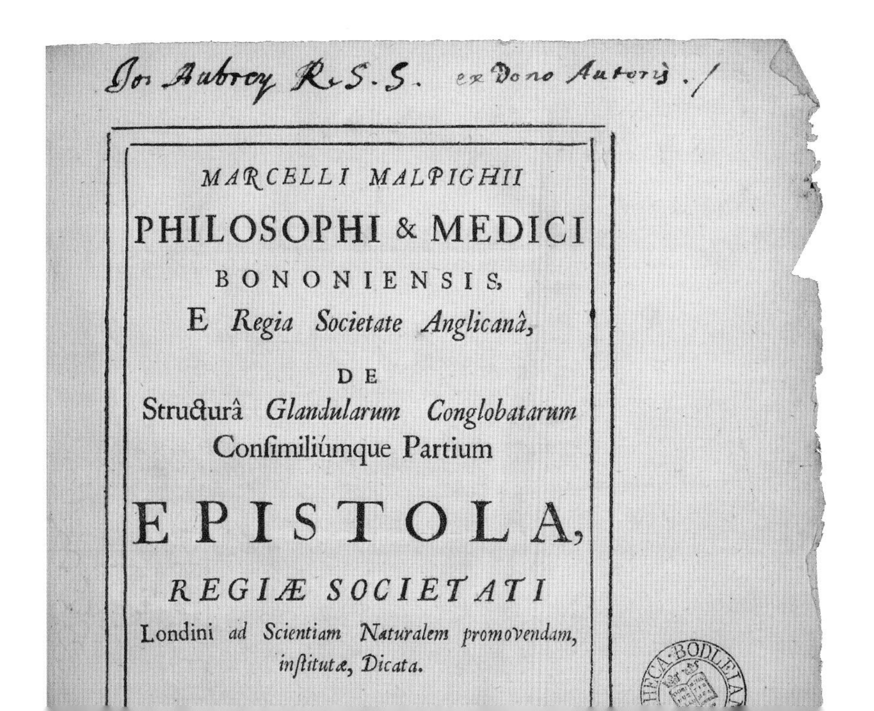

Jo: Aubrey R. S. S. ex Dono Autoris. /

MARCELLI MALPIGHII

PHILOSOPHI & MEDICI

BONONIENSIS,

E *Regia Societate Anglicanâ*,

DE

Structurâ *Glandularum Conglobatarum*

Consimiliúmque Partium

EPISTOLA,

*REGIÆ SOCIETATI*

Londini *ad Scientiam Naturalem promovendam, institutæ, Dicata.*

**Figure 73** Many of Aubrey's books donated to the Ashmolean were themselves gifts from their authors, for example Marcello Malpighi's *De structura glandularum Conglobatarum … epistola* (London 1689). Oxford, Bodleian Library, Ashmole 1692, title page.

# Select Bibliography

Abbreviations in the text above, and explained below: *BL* = Aubrey, *Brief Lives*, preferred edition below; *HP* = *The Hartlib Papers*; *PT* = *Philosophical Transactions*; *TPR* = *Three Prose Works*.

## Aubrey's Works

### Manuscripts

*Oxford, Bodleian Library*


MS Ashmole 1814, contains correspondence of Aubrey to Edward Lhwyd.

MSS Aubrey 1–2, 'The Naturall Historie of Wiltshire', parts 1–2.

MS Aubrey 3, 'An Essay towards a Description of the North Division of Wiltshire'.

MS Aubrey 4, 'A Perambulation of Surrey'.

MS Aubrey 5, 'An Interpretation of Villare Anglicanum'.

MS Aubrey 6, *Σχεδιασματα*. 'Brief Lives' Part I.

MS Aubrey 7, 'Brief Lives' Part II, and miscellaneous papers.

MS Aubrey 8, 'Brief Lives' Part III, and 'An Apparatus for the Lives of our English Mathematicians'.

MS Aubrey 9, 'Life of Thomas Hobbes of Malmesbury'.

MS Aubrey 10, 'An Idea of Education of Young Gentlemen and Idea Filioli seu Educatio Pueri', also containing miscellaneous papers and letters.

MS Aubrey 11, 'An Extract or Summary of the Lemmata of Stone-heng restored to the Danes' (by Walter Charleton).

MSS Aubrey 12–13, Letters to Aubrey A-N, P-Y.

MS Aubrey 16, copy of part of Aubrey's 'Chronologia Architectonica'.

MS Aubrey 17, 'Designatio de Easton-Piers in Comitatu Wilts'.

MS Aubrey 21, 'The Countrey Revell, or the Revell at Aldford', and miscellaneous papers, mainly in Aubrey's hand.

MS Aubrey 23, 'Collectio Geniturarum'.

MS Aubrey 24, 'Zercobeni, seu Claviculæ Salmonis Libri IV', transcribed with additions by Aubrey.

MS Aubrey 25, Nicolaus Mercator, *Musica*.

MS Aubrey 26, 'Faber Fortunæ'.

MS Ballard 14, Letters of Aubrey and others to Anthony Wood and Arthur Charlett.

MS e Musæo 149, 'Matthew of Westminster', 'Flores Historiarum' to 1303, with additions to 1306, bearing a number of inscriptions by Aubrey and Wood.

MS Eng. Misc. c. 323, *The History of the Temples of the Ancient Celts*.

MS Hearne's Diaries 158–9, notes from 'Analecta Ro Plot' (1675–76), partially derived from Aubrey's 'Adversaria Physica'.

MS Tanner 25, Letters to Thomas Tanner and others.

MS Tanner 456a, Letters of Dugdale, Wood, Aubrey and others.

MSS Top. Gen. c. 24–25, 'Monumenta Britannica'.

MS Wood F 39, Letters of Aubrey to Anthony Wood.


*Oxford, Worcester College*


MS 4.9 Mathematical notes and collections.

MS 5.4 Mathematical notes and collections.

MS 5.5 Mathematical notes and collections.

*London, British Library*
MS Egerton 2231, correspondence with Wood and others (copies).
MS Lansdowne 231, 'Remains of Gentilisme and Judaisme'.
MS Sloane 859, Francis Lodwick's library catalogue.
MS Stowe 1059, Henry Spelman, 'Archaesmus Graphicus'.

*London, Royal Society*
MS 92, MS of 'Natural History of Wiltshire', copied by commission in 1691.
Royal Society MS Early Letters P

## Editions

*Brief Lives, chiefly of contemporaries, set down by John Aubrey, between the years 1669 and 1696*, ed. Andrew Clark. 2 vols. Oxford, 1898. [There is no adequate edition of Aubrey's lives; this is still the best.]
*Three Prose Works*, ed. J. Buchanan-Brown. London: Centaur Press, 1972.
*Monumenta Britannica*, 3 parts in 2 volumes, ed. John Fowles and annot. Rodney Legg. Sherborne: Dorset Publishing Company, 1980, 1982.
*The Natural History of Wiltshire*, ed. John Britton. Wiltshire: Wiltshire Topographical Society, 1847.
*Wiltshire: The Topographical Collections of John Aubrey*, ed. J.E. Jackson. Wiltshire Archaeological and Natural History Society 1, 1862.
*The Natural History and Antiquities of the County of Surrey*, 5 vols, ed. and rev. Richard Rawlinson. London, 1718–19.
*Aubrey and Education*, ed. J.E. Stephens. London: Routledge & Kegan Paul, 1972.
*Miscellanies*. London, 1696. Revised and corrected edn. London, 1721.
*Two Antiquaries: a selection from the correspondence of John Aubrey and Anthony Wood*. Ed. M.G. Balme. Edinburgh: Durham Academic Press, 2001.

## PRIMARY LITERATURE

Anderson, James. *Historical Essay Shewing that the Crown and Kingdom of Scotland is Imperial and Independent*. Edinburgh, 1705.
Bacon, Francis. *Twoo Bookes ... of the Proficience and Advancement of Learning*. London, 1605.
———. *The Historie of the Raigne of King Henry the Seventh*. Cambridge, 1998; originally London, 1622.
———. *Sylva Sylvarum/New Atlantis*. London, 1627.
Birch, Thomas. *The History of the Royal Society of London*, 4 vols. London, 1756–57.
[Bolton, Edmund.] *Nero Caesar, or Monarchie Depraved*. London, 1624.
Burnet, Thomas. *Telluris theoria sacra/The Theory of the Earth*. London, 1681–89.
Camden, William. *Britannia*, ed. Edmund Gibson. London, 1695.
Carpenter, Nathanael. *Geographie Delineated*. Oxford, 1625.
Charleton, Walter. *Chorea Gigantum*. London, 1663.
Dalgarno, George. *Ars signorum*. London, 1661.
Daniel, Samuel. *Musophilus*, in *Poeticall Essayes*. London, 1599.
Evelyn, John. *The Diary of John Evelyn*, 6 vols. Oxford, 1955.
Gilbert, William. *De magnete, magneticisque morporibus, et de magno magnete tellure, physiologia Nova*. London, 1600.

Godwin, Francis. *The Man in the Moone*. London, 1638.

Greaves, John. *Pyramidographia, or a Description of the Pyramids in Ægypt*. Oxford, 1646.

Grew, Nehemiah. *Musæum Regalis Societatis*. London, 1681.

Halley, Edmond. 'An Account of the Cause of the Change of the Variation of the Magnetic Needle, with an Hypothesis of the Structure of the Internal Parts of the Earth.' *Philosophical Transactions* 16 (1692): 563–87.

Hartlib, Samuel. *Considerations tending to the Happy Accomplishment of Englands Reformation*. London, 1647.

*The Hartlib Papers*, 2nd edn, ed. Judith Crawford et al. 2 CD-ROMs. Ann Arbor, MI: UMI Research Publication, 2002.

Hearne, Thomas. *Peter Landtoft's Chronicle* (Oxford, 1725)

———. *Remarks and Collections*, 11 vols. Oxford: Clarendon Press for the Oxford Historical Society, 1885–1921.

Holinshed, Raphael. *The First and Second Volumes of Chronicles*. London, 1587.

Hooke, Robert. *Micrographia*. London, 1665.

———. *A Description of Helioscopes*. London, 1676.

———. *Posthumous Works*. Ed. Richard Waller. London: for the editor, 1705.

———. *The Diary of Robert Hooke M.A., M.D., F.R.S. 1672–80*. London: Taylor & Francis, 1935.

Jones, Inigo. *The Most Notable Antiquity of Britain, vulgarly called Stone-Heng, on Salisbury Plain*. London, 1655.

Kircher, Athanasius. *Mundus Subteranneus*. Amsterdam, 1665.

Kynaston, Francis. *The Constitutions of Musaeum Minervae*. London, 1636.

Lhuyd, Edward. *Lithophilacii Britannica ichnographia*. Oxford, 1699.

Lodwick, Francis. *A Common Writing*. London, 1647.

———. *The Ground-Work*. London, 1652.

Madox, Thomas. *Formulare Anglicanum*. London, 1702.

Martin, Martin. *A Escription of the Western Islands of Scotland*. London, 1703.

Osborne, Francis. *Advice to a Son*. Oxford, 1656.

Oughtred, William. *Clavis mathematicæ*. London, 1631; 4th edn 1667.

Petty, William. *The Advice of W.P.* London, 1647.

———. *The Discourse … concerning the Use of Duplicate Proportion*. London, 1674.

Plot, Robert. *The Natural History of Oxford-Shire*. Oxford, 1667.

———. *De origine fontium*. Oxford, 1685.

Sprat, Thomas. *The History of the Royal-Society of London*. London, 1667.

Steno, Nicolaus. *Prodromus*, trans. Henry Oldenburg. London, 1671.

Stillingfleet, Edward. *A defence of the Discourse concerning the Idolatry Practised in the Church of Rome*. London, 1676.

Stukeley, William. *Stonehenge, a Temple restor'd to the British Druids*. London, 1740.

———. *Abury, a Temple of the British Druids*. London, 1743.

Twining, Thomas. *Avebury in Wiltshire*. London, 1723.

Tradescant, John. *Musæum Tradescantianum*. London, 1656.

Wallis, John. *A Treatise of Algebra, both Historical and Practical*. London, 1685.

[Ward, Seth and John Wilkins.] *Vindiciæ Academiarum*. Oxford, 1654.

Webb, John. *A Vindication of Stone-Heng Restored*. London, 1665.

Webster, John. *Academiarum Examen, or the examination of academies*. London, 1654.

Whiston, William. *A New Theory of the Earth from the Original to the Consummation of Things*. London, 1696.

Wilkins, John. *The Discovery of a World in the Moone*. London, 1638.
———. *A Discourse concerning a New World, and Another Planet*. London, 1640.
———. *Mercury, or the Secret and Swift Messenger*. London, 1641.
———. *Mathematicall Magick*. London, 1648.
———. *An Essay towards a Real Character, and a Philosophical Language*. London, 1668.
Willis, Thomas. *Diatribæ duæ medico-philosophicæ*. London, 1659.
Wood, Anthony. *Athenæ Oxonienses*, 2 vols. Oxford, 1692–93. Rev. edn London, 1817.
Woodward, John. *An Essay towards a Natural History of the Earth*. London, 1695.
Wormius, Olaus. *Musæum Wormianum*. Leiden, 1655.
Wren, Stephen, ed. *Parentalia*. London, 1750.

## Secondary Literature

Bennett, J.A., Stephen Johnston and A.V. Simcock, *Solomon's House in Oxford: New Finds from the First Museum*. Oxford: MHS, 2000.
Bennett, J.A., and Scott Mandelbrote. *The Garden, the Ark, the Tower, the Temple: Biblical Metaphors of Knowledge in Early Modern Europe*. Oxford: Bodleian Library, 1998.
Bennett, Kate. 'Material towards a Critical Edition of John Aubrey's "Brief Lives".' University of Oxford, D.Phil. thesis, 1993.
———. 'Editing Aubrey.' In *Ma(r)king the Text*, ed. Joe Bray, Miriam Handley and Anne C. Henry. Aldershot: Ashgate, 2000, 271–90.
———. 'John Aubrey's Collections and the Early-Modern Museum.' *Bodleian Library Record* 17 (2001): 213–45.
———. 'John Aubrey, Hint-keeper: Life-writing and the Encouragement of Natural Philosophy in the pre-Newtonian Seventeenth Century.' *The Seventeenth Century* 22 (2007): 358–80.
———. 'John Aubrey and the "Lives of our English Mathematical Writers".' In *The Oxford Handbook of the History of Mathematics*, ed. Eleanor Robson and Jacqueline Stedall. Oxford: Oxford University Press, 2009, 329–52.
Buchanan-Brown, John. 'The Books Presented to the Royal Society by John Aubrey, F.R.S.' *Notes and Records of the Royal Society of London* 28 (1974): 167f.
Burl, Aubrey. *Prehistoric Avebury*. New Haven: Yale University Press 1979.
Cram, David. 'Universal Language, Specious Arithmetic and the Alphabet of Simple Notions.' *Beiträge zur Geschichte der Sprachwissenschaft* 4 (1994): 213–33.
Cram, David and Jaap Maat, eds. *George Dalgarno on Universal Language*. Oxford: Oxford University Press, 2001.
Emery, F.V. 'English Regional Studies from Aubrey to Defoe.' *Geographical Journal* 124 (1958): 315–25.
Feingold, Mordechai. The *Mathematician's Apprenticeship: Science, Universities and Society in England 1560–1640*. Cambridge: Cambridge University Press, 1984.
———. 'The Humanities', 'The Mathematical Sciences and New Philosophies', 'Oriental Studies' in *The History of the University of Oxford IV: The Seventeenth Century*, ed. Nicholas Tyacke. Oxford: Clarendon Press, 1997, 211–503.
———. 'The Origins of the Royal Society Revisited.' In *The Practice of Reform in Health, Medicine and Science 1500–2000*, ed. Margaret Pelling and Scott Mandelbrote, Aldershot: Ashgate, 2005, 167–83.

Fellows-Jensen, Gilliam. 'John Aubrey, Pioneer Onomast?' *Nomina* 23 (2000): 89–106.

Fox, Adam. 'Aubrey, John (1626–1697).' *Oxford Dictionary of National Biography*.

Frank, Robert G. 'John Aubrey, F.R.S., John Lydall, and Science at Commonwealth Oxford.' *Notes and Records of the Royal Society of London* 27 (1973): 193–217.

———. *Harvey and the Oxford Physiologists*. Berkeley: University of California Press, 1980.

Gordon, Cosmo A. 'Letter to John Aubrey from George Garden.' *Scottish Gaelic Studies* 8 (1955): 18–26.

Gunter, R.T. 'The Ashmolean Copy of Plot's "Natural History".' *Bodleian Library Quarterly* 6 (1930): 165–6.

———. 'The Library of John Aubrey, F.R.S.' *Bodleian Quarterly Record* 6 (1931): 230–36. Reprinted in Anthony Powell, *John Aubrey and His Friends*, London: Heinemann, 1963, Appendix B.

Haycock, David Boyd. *William Stukeley: Science, Religion and Archaeology in Eighteenth-Century England*. Woodbridge: Boydell Press, 2002.

Hazelgrove, D., and J. Murray. *John Dwight's Fulham Pottery 1672–1978*, published as *Journal of Ceramic History* 11 (1979).

Henderson, Felicity, and William Poole, eds. *Francis Lodwick: Writings on Language, Theology, and Utopia*. Oxford: Oxford University Press, 2010.

Hoppen, K.T. *The Common Scientist in the Seventeenth Century: A Study of the Dublin Philosophical Society 1683–1708*. London: Routledge & Kegan Paul, 1970.

Hunter, Michael. 'The Royal Society and the Origins of British Archaeology.' *Antiquity* 65 (1971): 113–21, 187–92.

———. 'The Bibliography of John Aubrey's *Brief Lives*.' *Antiquarian Book Monthly Review* 1 (1974): 6f.

———. *John Aubrey and the Realm of Learning*. London: Duckworth, 1975.

———. *Science and Society in Restoration England*. Cambridge: Cambridge University Press, 1981.

———. *The Royal Society and its Fellows, 1660–1700: The Morphology of an Early Scientific Institution*, 2nd edn. Chalfont St Giles: BSHS, 1994.

Hutton, Ronald. *Blood and Mistletoe: The History of the Druids in Britain*. New Haven: Yale University Press, 2009.

Impey, O.R., and Arthur MacGregor, eds. *The Origins of Museums*. London: House of Stratus, 2001.

Kelly, Joseph. 'A Book from the Libraries of Ben Jonson and John Aubrey.' *Seventeenth-Century News* 39 (1981): 44.

Kollerstrom, Nicholas. 'The Hollow World of Edmund Halley.' *Journal for the History of Astronomy* 23 (1992): 185–92.

Lewis, Rhodri. 'The Efforts of the Aubrey Correspondence Group to Revise John Wilkins's *Essay* (1668) and Their Context.' *Historiographia Linguistica* 28 (2001): 333–66.

———. *Language, Mind and Nature: Artificial Languages in England from Bacon to Locke*. Cambridge: Cambridge University Press, 2007.

MacGregor, Arthur G. *The Ashmolean Museum*. Oxford: Ashmolean Museum, 2001.

———. 'The Antiquary *en plain air*: Eighteenth-century Progress from Topographical Survey to the Threshold of Field Archaeology.' In R.G.W. Anderson, Marjorie Lancaster Caygill, Arthur G. MacGregor and Luke Syson, eds, *Enlightening the British: Knowledge, Discovery and the Museum in the Eighteenth Century*. London: British Museum Press, 2003, 164–75.

Malcolm, Noel and Jacqueline Stedall. *John Pell (1611–1685) and His Correspondence with Sir Charles Cavendish: The Mental World of an Early Modern Mathematician*. Oxford: Oxford University Press, 2005.

Moorhead, T.S.N. 'A Roman Coin Hoard from Wanborough.' In A.S. Anderson, J.S. Wacher and A.P. Fitzpatrick, *The Romano-British 'Small Town' at Wanborough, Wiltshire*, Appendix 1. Britannia Monograph Series no. 19. London: Society for the Promotion of Roman Studies, 2001.

Parry, Graham. *The Trophies of Time: English Antiquarians of the Seventeenth Century*. Oxford: Oxford University Press, 1995.

Piggott, Stuart. *William Stukeley: An Eighteenth-century Antiquary*, rev. edn. London: Thames & Hudson, 1985.

———. *Ancient Britons and the Antiquarian Imagination*. London: Thames & Hudson, 1989.

Poole, William. 'The Genesis Narrative in the Circle of Robert Hooke and Francis Lodwick', in *Scripture and Scholarship in Early Modern England*, ed. A. Hessayon and N. Keene. Aldershot: Ashgate, 2006, 41–57.

———. 'Sir Robert Southwell's Dialogue on Thomas Burnet's Theory of the Earth: "C & S discourse of Mr Burnetts Theory of the Earth" (1684).' *The Seventeenth Century* 23 (2008): 72–104.

———. *The World Makers: Scientists of the Restoration and the Search for the Origins of the Earth*. Oxford: Peter Lang, 2010.

Powell, Anthony. *John Aubrey and His Friends*. London: Eyre & Spottiswoode, 1948. Revised edition, London: Heinemann, 1963.

———. 'John Aubrey's Books I, II.' *Times Literary Supplement* 13, 20 January 1950, pp. 32, 48.

Rudwick, Martin J.S. *The Meaning of Fossils: Episodes in the History of Palaeontology*, 2nd edn. Chicago: University of Chicago Press, 1985.

Seaton, Ethel. *Literary Relations of England and Scandinavia in the Seventeenth Century*. Oxford: Clarendon Press, 1935.

Shalev, Zur. 'Measurer of All Things: John Greaves (1602–1652), the Great Pyramid, and Early Modern Metrology.' *Journal of the History of Ideas* 63 (2002): 555–75.

Stedall, Jacqueline A. *A Discourse concerning Algebra: English Algebra to 1685*. Oxford: Oxford University Press, 2002.

Syfret, R.H. 'Some Early Reactions to the Royal Society.' *Notes and Records of the Royal Society of London* 7 (1950), 207–58.

Turnbull, G.H. 'Samuel Hartlib's Acquaintance with John Aubrey.' *Notes & Queries* 195 (1950): 31–3.

———. 'Samuel Hartlib's Connection with Sir Francis Kynaston's Musæum Minervæ.' *Notes & Queries* 197 (1952): 33–37.

Tylden-Wright, David. *John Aubrey: A Life*. London: HarperCollins, 1991.

Turner, A.J. 'Mathematical Instruments and the Education of Gentlemen.' *Annals of Science* 30 (1973): 51–88.

———. 'Andrew Paschall's Tables of Plants for the Universal Language, 1678.' *Bodleian Library Record* 9 (1978): 346–50.

Ucko, Peter J., et al. *Avebury Reconsidered: From the 1660s to the 1990s*. London: Unwin Hyman, 1991.

Webster, Charles. *The Great Instauration: Science, Medicine and Reform 1626–1660*, 2nd edn. Oxford: Peter Lang, 2002.